Numbers and Shapes Revisited

Numbers and Shapes Revisited
More Problems for Young Mathematicians

JUDITA COFMAN

CLARENDON PRESS · OXFORD
1995

Oxford University Press, Walton Street, Oxford OX2 6DP

Oxford   New York
Athens   Auckland   Bangkok   Bombay
Calcutta   Cape Town   Dar es Salaam   Delhi
Florence   Hong Kong   Istanbul   Karachi
Kuala Lumpur   Madras   Madrid   Melbourne
Mexico City   Nairobi   Paris   Singapore
Taipei   Tokyo   Toronto
and associated companies in
Berlin   Ibadan

Oxford is a trade mark of Oxford University Press

Published in the United States
by Oxford University Press Inc., New York

A catalogue record for this book is available from the British Library

Library of Congress Cataloging in Publication Data
Cofman, Judita.
Numbers and shapes revisited: more problems for young mathematicians/Judita Cofman.
Includes bibliographical references and index.
1. Problem solving.   2. Mathematics—Problems, exercises, etc.
I. Title.
QA63. C63   1995   510—dc20   94-27279
ISBN 0 19 853460 4

Typeset by Alden Multimedia, Northampton
Printed in Great Britain by
Biddles Ltd, Guildford and King's Lynn

Preface

Mathematics syllabuses in modern secondary schools contain a large amount of notions and basic facts from various mathematical disciplines. There is usually no time in lessons for explaining details and pointing out connections between parts of the curriculum. Pupils with a keen interest in the subject should be encouraged to expand their knowledge and to explore links between mathematical phenomena. *Numbers and shapes revisited* is an attempt to assist the pursuit of such studies.

The aims of the book are:

(a) To acquaint the reader with a selection of classical topics, ranging from properties of numbers, congruences, inequalities, shapes in two- and higher-dimensional spaces to applications of groups and problems of combinatorics.

(b) To encourage the search for interrelations between objects, facts and ideas encountered at work. Each chapter in the book contains topics which are related one to another in some ways. For example, in Chapter I patterns of generalized Fibonacci numbers are compared with patterns of polygons, in Chapter VII triangles are compared with tetrahedra as their three-dimensional analogues, and in Chapter IX groups are applied for solution of functional equations.

(c) To foster independent thinking. For this reason the material is presented in the form of problems for solution. In Part 1 of the book each chapter starts with definitions and fundamental properties of the relevant concepts. Then come the statements of the problems with explanations and occasional hints. This approach to study should give the reader the chance to stop at crucial points marked by the questions raised, to reflect on their relevance to the topics considered, and to try to produce some answers. Detailed solutions to the problems are described in Part 2.

The problems differ in nature: some concern proofs of well known facts, while others are technical exercises.

Suggestions for using the book

It could be helpful to regard *Numbers and shapes revisited* as a sequel to *What to solve?*. Both books are based on work at International Camps for

Young mathematicians. There the youngsters first polished their problem-solving skills by answering questions similar to those in *What to solve?*. After that they concentrated on topics described in *Numbers and shapes revisited*. A similar approach is recommended to the reader: preliminary study of approaches to problem solving, discussed in *What to solve?* should facilitate the reading of *Numbers and shapes revisited*. Besides, *What to solve?* contains appendices with useful details, such as basic definitions, biographical data about mathematicians and an extensive bibliography, which are not repeated in the second book.

Each chapter in the book can be studied independently of the others (although there are a few cross-references between some chapters). The answers to the questions, given in part 2, are not necessarily the unique solutions to the problems in the book.

The book ends with lists of references for each chapter, containing suggestions for further reading.

Who could use the book?

The book is intended for advanced secondary school pupils, aged fifteen and over, but it is hoped that undergraduates, students at teachers training colleges, and mathematics teachers can also benefit from it.

Erlangen
March 1994   J.C.

vi

Acknowledgements

I would like to express my gratitude to Abe Shenitzer for his continuous encouragement and invaluable advice during the preparation of this book.

I am deeply indebted to Terry Heard who read the manuscript and suggested a number of improvements.

My thanks go to Helen Hodgson for the beautiful typing.

The manuscript was prepared while I was working at Putney High School, London. I wish to thank the Headmistress Eileen Merchant, and my former colleagues, especially Rosemary Gooding. Without their friendly support the book would never have been written. Sincere thanks are due to Oxford University Press for publishing the book.

Contents

Part 2   Solutions

Part 1
Problems

I The Fibonacci sequence, generalized Fibonacci sequences, and related topics

Introduction

One of the most engaging puzzles in the history of mathematics was formulated by Leonardo of Pisa, nicknamed Fibonacci (1170–1250) in his celebrated book *Liber abaci*:

> How many pairs of rabbits will be produced in a year, beginning with a single pair, if in every month each pair bears a new pair which becomes productive from the second month on?

It is left to the reader to verify that the numbers of pairs of rabbits in consecutive months form the sequence F:

$$1, 1, 2, 3, 5, 8, 13, \ldots.$$

The sequence F is nowadays called the *Fibonacci sequence*; its terms are the Fibonacci numbers f_n. It is customary to denote the initial term of the Fibonacci sequence by f_0, the next term by f_1, the following term by f_2, and so on.

In the Fibonacci sequence each term, starting from f_2, is the sum of the two previous terms. Thus:

The Fibonacci numbers can be defined by the following recursion formula:

$$\left.\begin{array}{l} f_n = f_{n-1} + f_{n-2} \quad \text{for } n \geqslant 2 \\ f_1 = f_0 = 1. \end{array}\right\} \tag{1.1}$$

The Fibonacci sequence is not only the solution of an intriguing puzzle. Fibonacci would be surprised to learn how important 'his' numbers turned out to be in mathematics, in science, and in art!

Mathematicians like to play with ideas. After a thorough study of the Fibonacci sequence the following question emerged.

If one altered the recursion formula in various ways, how would the alterations affect the properties of the resulting number sequences?

Diverse generalizations of the Fibonacci sequence F were suggested, leading to collections of sequences that included F as a special case.

3

Our aims in this chapter are:

(a) To present a selection of problems concerned with elementary properties of the Fibonacci numbers (see Section 1.2).

(b) To introduce three types of generalizations of F prompting the study of related topics. These generalizations are described below.

   (I) The k-bonacci sequence $F(k)$ consisting of the terms $f_0(k), f_1(k), \ldots, f_n(k), \ldots$ is defined for any natural number $k \geqslant 2$ by the recursion formula

   $$\left.\begin{array}{ll} f_n(k) = f_{n-1}(k) + f_{n-2}(k) + \ldots + f_{n-k}(k) & \text{for } n \geqslant k \\ f_n(k) = f_{n-1}(k) + f_{n-2}(k) + \ldots + f_0(k) & \text{for } 1 \leqslant n < k \\ f_o(k) = 1. \end{array}\right\}$$

   $$(1.2)$$

   $F(k)$ generalizes F, since $F(2) = F$. The k-bonacci sequence will lead us to generalizations of the famous *golden rectangle* (Section 1.3).

   (II) $F^{(i)}$, *the Fibonacci sequence of order* i, with terms $f_0^{(i)}$, $f_1^{(i)}$, $f_2^{(i)}, \ldots, f_n^{(i)}, \ldots$ is defined for $i = 1, 2, 3, \ldots$ recursively with the help of $F^{(i-1)}$, that is of the Fibonacci sequence of order $i-1$, as follows:

   $$\left.\begin{array}{ll} f_n^{(i)} = f_{n-1}^{(i)} + f_{n-2}^{(i)} + f_n^{(i-1)} & \text{for } n \geqslant 2, \ i \geqslant 1 \\ f_0^{(i)} = 1 \text{ and } f_1^{(i)} = i+1 & \text{for } i > 0 \\ f_n^{(0)} = f_n \text{ for } n \geqslant 0. \end{array}\right\}$$

   $$(1.3)$$

   Again, $F^{(i)}$ is a generalization of F, because $F^{(0)} = F$. The study of $F^{(i)}$ will lead to the construction of an intriguing fractal, recommended to computer fans (Section 1.4).

   (III) Let a and b be arbitrary numbers. The *sequence* $F(a, b)$ with terms $f_0(a, b), f_1(a, b), \ldots, f_n(a, b), \ldots$ is defined by the following alteration of (1.1):

   $$\left.\begin{array}{ll} f_n(a, b) = af_{n-1}(a, b) + bf_{n-2}(a, b) & \text{for } n \geqslant 2 \\ f_0(a, b) \text{ and } f_1(a, b) \text{ are given numbers.} \end{array}\right\}$$

   $$(1.4)$$

   Obviously, $F = F(1, 1)$ with $f_0(1, 1) = f_1(1, 1) = 1$. In Section 1.5 we shall show that $f_n(a, b)$ can be expressed in terms of a, b, and n without referring to previous terms of the sequence.

We begin with some puzzles.

1.1   A few puzzles on house painting

The following 'colourful' puzzle is taken from Rényi [3]:

Problem 1

On a housing estate a house with n floors has to be painted under the following restrictions:

(1)   each floor of the house has to be painted either white or blue;

(2)   no two blue floors can be directly on top of one another.

Find out: In how many different ways can an n-floored house be painted under the above restrictions for any given natural number n?

By altering the painting rules new problems can be created. Here are a few variations.

Problem 2

The floors of a house are painted white or blue so that no three blue floors can be directly on top of one another. In how many different ways can a house with n floors be painted under these restrictions?

Problem 3

Each floor of a house has to be painted white, blue, or grey, and no two blue floors can be on top of one another. Find a recurrence relation for the number of distinct ways of painting a house with n floors.

Problem 4

Find a recurrence relation for the total number of blue floors in the set of all differently painted houses with n floors subject to rules (1) and (2) of Problem 1. (For example, the total number of blue floors for the three possible ways of painting of a 2-floored house is 2.)

1.2   Some well-known properties of the Fibonacci numbers

Problem 5

Prove the following identities, satisfied by the Fibonacci numbers:

(a) $$f_0 + f_1 + f_2 + \ldots + f_{n-1} = f_{n+1} - 1 \qquad \text{for } n \geqslant 1;$$

(b) $$f_0 + f_2 + f_4 + \ldots + f_{2n} = f_{2n+1} \qquad \text{for } n \geqslant 0;$$

(c) $\qquad f_1 + f_3 + f_5 + \ldots + f_{2n-1} = f_{2n} - 1 \quad$ for $n \geqslant 1$;

(d) for $n \geqslant 1$ the successive Fibonacci numbers f_n and f_{n+1} are co-prime (that is, their highest common factor is 1).

Problem 6

(a) Prove that
$$f_{n+m} = f_{n-1} f_{m-1} + f_n f_m \quad \text{for } n \geqslant 1, m \geqslant 1.$$

(b) Deduce that if n is divisible by m then f_{m-1} divides f_{n-1}.

Problem 7

(a) Prove that for $n \geqslant 2$ the nth power of the 2×2 matrix $A = \begin{pmatrix} 1 & 1 \\ 1 & 0 \end{pmatrix}$ is

$$A^n = \begin{pmatrix} f_n & f_{n-1} \\ f_{n-1} & f_{n-2} \end{pmatrix}. \tag{1.5}$$

(b) Using (a) find the value of
$$f_{n+1} \cdot f_{n-1} - f_n^2 \quad \text{for } n \geqslant 1.$$

Problem 8

(a) Prove the identity, known as Binet's formula:

$$f_n = \frac{1}{\sqrt{5}} \left[\left(\frac{1 + \sqrt{5}}{2} \right)^{n+1} - \left(\frac{1 - \sqrt{5}}{2} \right)^{n+1} \right] \quad \text{for } n \geqslant 0 \tag{1.6}$$

(Binet (1786–1856) was a French mathematician.)

(b) Deduce that

$$\lim_{n \to \infty} \frac{f_n}{f_{n-1}} = \frac{1 + \sqrt{5}}{2} \tag{1.7}$$

(that is, for large values of n the ratio f_n/f_{n-1} approaches $(1 + \sqrt{5})/2$).

1.3 k-bonacci numbers and generalizations of the golden rectangle

The Fibonacci numbers often appear in geometry as lengths in different geometrical figures. A simple example is a rectangle with side lengths f_n and f_{n-1} for any $n \geqslant 1$. Call this rectangle the nth *Fibonacci rectangle* R_n. (Figure 1.1 shows the rectangles R_1, R_2, R_3, and R_4.)

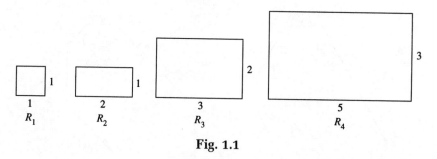

Fig. 1.1

It is not difficult to verify the following:

Problem 9

(a) No two rectangles of the sequence $R_1, R_2, R_3, \ldots$ are similar to one another; however

(b) for large values of n the shape of R_n approaches the shape of a rectangle R_G whose side lengths are in the ratio $(1 + \sqrt{5})/2 : 1$.

The beauty of this statement is enhanced by the fact that the shape which R_n approaches for large values of n is the famous shape of the *golden rectangle*. The golden rectangle was appreciated already by the ancient Greeks and since then it has continued to delight mathematicians and artists.

By definition,

The golden rectangle is a rectangle whose side lengths h and b satisfy the proportion $h : b = b : (h + b)$.

It is left to the reader to deduce that $b/h = (1 + \sqrt{5})/2$. Hence $(1 + \sqrt{5})/2$ is called the *golden ratio*; it is usually denoted by the Greek letter φ (pronounced phi).

In his book *More mathematical puzzles and diversions,* [1] Martin Gardner tells us that in around 1913 Mark Barr generalized the concept of the golden ratio by proving that:

For any $k \geqslant 2$ in the k-bonacci sequence the ratio $f_n(k)/f_{n-1}(k)$ of consecutive terms approaches a limit — call it $\varphi(k)$ — as n tends to infinity ($\varphi(2) = \varphi$).

The proof of the above statement is beyond the scope of this book. We shall, however, accept its truth and use it to generalize the golden rectangle as follows:

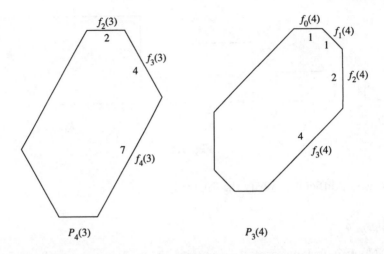

Fig. 1.2

The golden rectangle R_G is a four-sided polygon with equal angles and consecutive sides a_1, a_2, a_3, a_4 such that $a_2 : a_1 = \varphi(2)$ and $a_3 = a_1$, $a_4 = a_2$. We shall define an analogous shape $P_S(k)$ related to $\varphi(k)$:

> $P_S(k)$ *is a $2k$-sided polygon with equal angles and consecutive sides* $a_1, a_2, a_3, \ldots, a_k, a_{k+1}, a_{k+2}, \ldots, a_{2k}$ *such that* $a_2 : a_1 = a_3 : a_2 = \ldots$ $= a_k : a_{k-1} = \varphi(k)$, *and* $a_1 = a_{k+1}, a_2 = a_{k+2}, \ldots, a_k = a_{2k}$.

Let us call $P_S(k)$ *the silver $2k$-gon.* (The silver 4-gon $P_S(2)$ is the golden rectangle R_G.)

The Fibonacci rectangles R_n also admit generalizations related to the k-bonacci numbers:

> *Let us define the nth k-bonacci $2k$-gon as a $2k$-sided polygon with equal angles and consecutive side lengths* $f_n(k), f_{n-1}(k), f_{n-2}(k), \ldots, f_{n-(k-1)}(k),$ $f_n(k), f_{n-1}(k), \ldots, f_{n-(k-1)}(k)$. *This polygon will be denoted by* $P_n(k)$.

Figure 1.2 shows $P_4(3)$ and $P_3(4)$.

Since $f_n(k)/f_{n-1}(k)$ *converges to* $\varphi(k)$, *for large values of n the polygon* $P_n(k)$ *approaches the shape of the silver polygon* $P_S(k)$.

We can now return to problem solving, related to the newly introduced shapes. We shall compare some of their properties with those of R_n and R_G.

Problem 10

(a) Prove the well-known property of R_n, that for $n \geqslant 2$ it can be partitioned (that is, divided without gaps and overlapping) into two polygons, one of which is R_{n-1} and the other the 'Fibonacci square' S_{n-1} of side length f_{n-1}.

(b) Deduce, by comparing areas, the following identity for Fibonacci numbers:

$$f_n f_{n-1} = f_{n-1} f_{n-2} + f_{n-1}^2. \tag{1.8}$$

(c) Prove that R_n can be partitioned into the Fibonacci squares $S_{n-1}, S_{n-2}, \ldots, S_1, S_0$ and deduce that the product $f_n f_{n-1}$ is the sum of the square numbers:

$$f_n f_{n-1} = f_{n-1}^2 + f_{n-2}^2 + \ldots + f_0^2. \tag{1.9}$$

One wonders — can geometry help us to formulate generalizations of (1.8) and (1.9) for k-bonacci numbers? The answer to this question is affirmative. Let us first consider the case of 3-bonacci hexagons $P_n(3)$ (because they can be easily drawn and visualized).

Problem 11

(a) Prove that for $n \geqslant 3$ the 3-bonacci hexagon $P_n(3)$ can be partitioned into two polygons, one of which is $P_{n-1}(3)$ and the other a concave hexagon $P_{n-1}'(3)$.

(b) Deduce, by comparing areas, the following identity for 3-bonacci numbers:

$$f_n(3) f_{n-1}(3) + f_n(3) f_{n-2}(3) + f_{n-1}(3) f_{n-2}(3)$$
$$= f_{n-1}(3) f_{n-2}(3) + f_{n-1}(3) f_{n-3}(3) + f_{n-2}(3) f_{n-3}(3)$$
$$+ [f_{n-1}(3) + f_{n-2}(3)]^2. \tag{1.10}$$

(c) Prove that $P_n(3)$ can be partitioned into the concave hexagons $P_{n-1}'(3), P_{n-2}'(3), \ldots, P_2'(3)$ and the 3-bonacci hexagon $P_2(3)$. Deduce that

$$f_n(3) f_{n-1}(3) + f_n(3) f_{n-2}(3) + f_{n-1}(3) f_{n-2}(3)$$
$$= [f_{n-1}(3) + f_{n-2}(3)]^2 + [f_{n-2}(3) + f_{n-3}(3)]^2 + \ldots$$
$$+ [f_1(3) + f_0(3)]^2. \tag{1.11}$$

Problem 12

Comparison of (1.8) with (1.10) and of (1.9) with (1.11) suggests formulae generalizing (1.8) and (1.9) for arbitrary k-bonacci numbers. State them and try to prove them algebraically.

The next two problems are about the golden rectangle and the silver hexagon.

Problem 13

Figure 1.3 shows a sequence of nested golden rectangles:

$$(R_G)_0 = A_0 A_1 A_2 A_2'$$
$$(R_G)_1 = A_1 A_2 A_3 A_3'$$
$$(R_G)_2 = A_2 A_3 A_4 A_4'$$
$$\cdots\cdots\cdots\cdots\cdots$$
$$(R_G)_5 = A_5 A_6 A_7 A_7'.$$

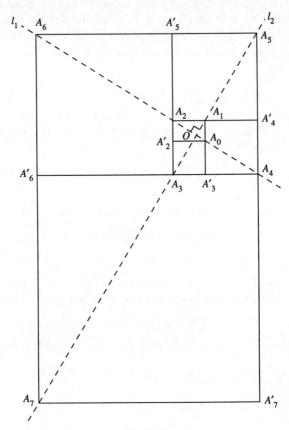

Fig. 1.3

Prove the following.

(a) The diagonals A_0A_2, A_1A_3, A_2A_4, A_3A_5, A_4A_6, and A_5A_7 of these golden rectangles are on two straight lines ℓ_1 and ℓ_2 meeting at right angles in a point O.

(b) The points A_0, A_1, A_2, A_3, A_4, A_5, A_6, and A_7 belong to a common logarithmic spiral σ with pole O. This logarithmic spiral is called the *golden spiral*. (*Hint.* Recall that if a straight line ℓ rotates uniformly about a fixed point O — called the pole — and if a point M moves along ℓ receding from O at a rate proportional to the distance OM, then the curve described by M is a *logarithmic spiral* (Fig. 1.4).)

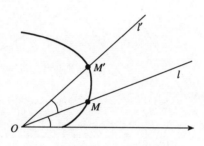

Fig. 1.4

Problem 14

Figure 1.5 depicts four nested silver hexagons: $(P_S(3))_0 = A_0A_1A_2A_3A_3'A_3''$, $(P_S(3))_1 = A_1A_2A_3A_4A_4'A_4''$, $(P_S(3))_2 = A_2A_3A_4A_5A_5'A_5''$, and $(P_S(3))_3 = A_3A_4A_5A_6A_6'A_6''$. Prove that:

(a) The diagonals A_0A_3, A_1A_4, A_2A_5, and A_3A_6 of these hexagons are on three straight lines ℓ_1, ℓ_2, ℓ_3 that meet at a common point O.

(b) The points A_0, A_1, A_2, A_3, A_4, A_5, and A_6 belong to a common logarithmic spiral $\sigma(3)$.

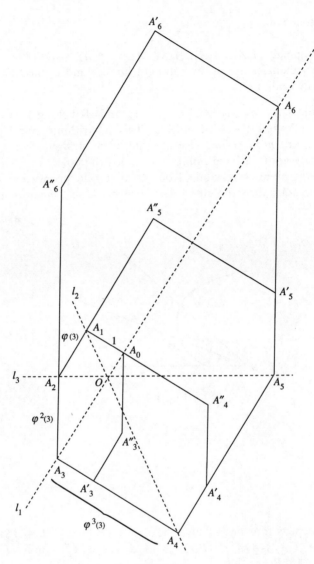

Fig. 1.5

We conclude this section with a problem involving a three dimensional generalization of the Fibonacci rectangle R_n, namely the *'3-bonacci cuboid'* $C_n(3)$, defined as a cuboid with side lengths $f_n(3)$, $f_{n-1}(3)$, and $f_{n-2}(3)$ where $n \geqslant 2$. The problem is given below.

Problem 15

By nesting 3-bonacci cuboids derive the formula

$$f_n(3)\,f_{n-1}(3)\,f_{n-2}(3) = 1 \cdot 1(1+1) + 1 \cdot 2(1+2) + \ldots$$
$$+ f_{n-2}(3)\,f_{n-1}(3)\,[\,f_{n-2}(3) + f_{n-1}(3)\,].$$

1.4 From Fibonacci numbers of higher order to fractals

The Fibonacci sequences $F^{(i)}$ or order i and their terms $f_n^{(i)}$ were introduced by means of the recurrence relations (1.3) on p. 3. In our Maths Club we decided to construct a number triangle, by listing the terms of $F^{(i)}$ in (vertical) columns for $i = 0,1,2,\ldots$, as shown in Fig. 1.6.

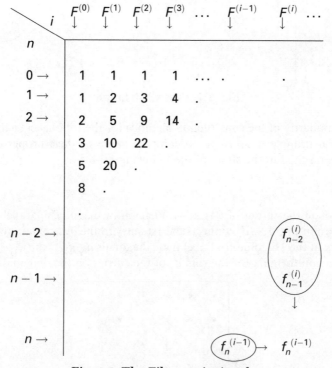

Fig. 1.6 The Fibonacci triangle

This idea was rewarding for the following reason: the construction of this number triangle — which we called the 'Fibonacci triangle' — reminded us of the construction of the famous Pascal triangle (Fig. 1.7). In the latter each entry (except for those in the 0th row and in the 0th column) is the sum of one entry on its left and *one* entry directly above it. In the Fibonacci triangle each entry (except those in the 0th column and in the 0th and first rows) is obtained by adding one entry on its left to *two* entries directly above it.

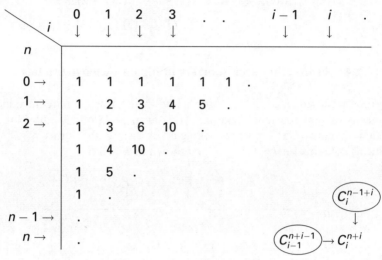

Fig. 1.7 Pascal's triangle

The similarity of the construction methods for the Fibonacci triangle and the Pascal triangle tempted us to compare some of their properties. Our attempts resulted in the study of the following problems.

Problem 16

(a) Through the entries of Pascal's triangle draw diagonals parallel to d_1 as shown in Fig. 1.8 (d_1 connects the 1st entry in the 0th column to the 0th entry in the 1st column). Label these diagonals $d_0, d_1, \ldots, d_n, \ldots$. Find a recursion formula for the sum $\bar{d}_n$ of the entries on the diagonal d_n.

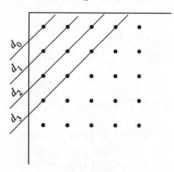

Fig. 1.8

(b) Through the entries of the Fibonacci triangle draw the same set of diagonals $d_0, d_1, \ldots, d_n, \ldots$. Find a recursion formula for the sum $\bar{d}'_n$ of the entries d_n.

(c) Through the entries of Pascal's triangle draw steeper diagonals $s_0, s_1, s_2, \ldots, s_n, \ldots$ parallel to s_2 as shown in Fig. 1.9 (s_2 connects the 2nd entry of the 0th column with the 0th entry of the 1st column.) Find a recursion formula for the sum $\bar{s}_n$ of the entries on s_n.

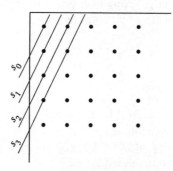

Fig. 1.9

(d) Through the entries of the Fibonacci triangle draw the same set of diagonals $s_1, s_2, \ldots, s_n \ldots$ and find a recursion formula for the sum $\bar{s}'_n$ of the entries on diagonal s_n.

Problem 17

(a) Consider Pascal's triangle. For each $i \geq 1$ and for each $n \geq 0$ the entry in the nth row of the ith column is the sum of the first n entries in the preceding $(i-1)$th column. Prove this assertion.

(b) Consider the Fibonacci triangle. For any $i \geq 1$ and for any $n \geq 0$ the entry in the nth row of the ith column can be expressed in terms of the entries of the preceding $(i-1)$th column and of the 0th column (that is, of the Fibonacci numbers) as follows:

$$f_n^{(i)} = f_0^{(0)} f_n^{(i-1)} + f_1^{(0)} f_{n-1}^{(i-1)} + \ldots + f_n^{(0)} f_0^{(i-1)}.$$

Prove this statement.

(c) Deduce from (b) that $f_n^{(i)}$ can be written as the sum of products of Fibonacci numbers taken $i+1$ at a time.

A brief glance at Pascal's triangle reveals that each of its diagonals d_n contains odd entries; moreover, some of its diagonals (such as d_0, d_1, d_3, d_7) involve just odd entries. Nevertheless, in a 'very large Pascal triangle', that is one in which many diagonals are listed, odd numbers occur with probability very close to zero. To gain an insight into the distribution of odd and even entries in Pascal's triangle we reduce its entries modulo 2:

Problem 18

If in Pascal's triangle each entry is replaced by its remainder when divided by 2, the resulting pattern of zeros and ones is called the *Pascal's triangle modulo 2*. Construct a 'large' copy of Pascal's triangle modulo 2. (This can be done easily with the help of a computer.)

Do the same for the Fibonacci triangle:

Problem 19

The *Fibonacci triangle modulo 2* is the number pattern obtained from the Fibonacci triangle by replacing its odd entries by 1 and its even entries by 0. Construct a 'large' copy of the Fibonacci triangle modulo 2. (Again, the use of a computer is recommended.)

The triangles of zeros and ones can be turned into beautiful pictures if the ones are replaced by small black squares and the zeros by white squares of the same size, as shown below (Fig. 1.10) for a fragment of Pascal's triangle:

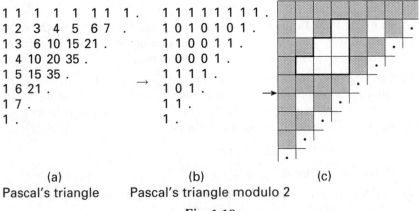

(a)	(b)	(c)
Pascal's triangle	Pascal's triangle modulo 2	

Fig. 1.10

Mathematicians were surprised to find out that the pattern of the black and white squares obtained from a 'large' copy of Pascal's triangle modulo 2 resembles the so-called *Sierpiński gasket,* nowadays recognized as a fractal. (The study of fractals was initiated by B. Mandelbrot in the second half of this century. The Sierpiński gasket, named after the renowned Polish mathematician W. Sierpiński (1882–1969), was constructed earlier.)

It is an aesthetic pleasure to look at the Sierpiński gasket, and the study of its geometrical structure is fascinating. The corresponding 'Fibonacci gasket' obtained from the Fibonacci triangle modulo 2 is even more intriguing. Readers are encouraged to employ their computers and to attack the last problem in this section.

Problem 20

(a) In a 'large' copy of Pascal's triangle modulo 2 replace the ones and the zeros by black and white squares respectively. You will notice a pattern of black and white triangles in the picture (one white triangle, consisting of six white squares, can be seen even in the small fragment in Fig. 1.10(c)). Try to analyse this pattern.

(b) Do the same for a 'large' copy of the Fibonacci triangle modulo 2.

1.5 How to solve recurrence relations of the form $f_n(a, b) = af_{n-1}(a, b) + bf_{n-2}(a, b)$

In Problem 8 the reader was asked to verify Binet's formula (1.6), in which f_n is expressed in terms of n only without referring to preceding terms of the Fibonacci sequence F. It is not difficult to verify the validity of (1.6) but it is natural to ask how it is derived.

The aim of this section is to answer this question not only for the Fibonacci numbers, but for all generalized Fibonacci numbers $f_n(a, b)$ satisfying the recurrence relation

$$f_n(a, b) = af_{n-1}(a, b) + bf_{n-2}(a, b) \quad \text{for } n \geqslant 2 \qquad (1.12)$$

for given values of a and b, and given intial terms $f_0(a, b)$ and $f_1(a, b)$.

We shall present a method enabling us to express $f_n(a, b)$ in terms of $a, b,$ and n only. This will be done through a series of problems.

Problem 21

If two number sequences

$$g_0, g_1, g_2, \ldots, g_n, \ldots \quad \text{and} \quad h_0, h_1, h_2, \ldots, h_n, \ldots$$

satisfy the same recurrence relation (1.12) then, for arbitrary not necessarily distinct numbers A and B, the sequence

$$Ag_0 + Bh_0, \; Ag_1 + Bh_1, \ldots, \; Ag_n + Bh_n, \ldots$$

also satisfies (1.12). Prove this statement.

There is a most useful link between the recurrence relation (1.12) and the quadratic equation $x^2 = ax + b$:

Problem 22

(a) If α is a solution of the equation $x^2 = ax + b$, prove that the sequence of the powers of α.

$$\alpha^0, \alpha^1, \alpha^2, \alpha^3, \ldots, \alpha^n, \ldots$$

satisfies (1.12).

(b) Deduce that if α and β are two *distinct* roots of the equation $x^2 = ax + b$, then for two arbitrary not necessarily distinct numbers A, B the sequence

$$A\alpha^0 + B\beta^0, A\alpha^1 + B\beta^1, \ldots, A\alpha^n + B\beta^n, \ldots$$

satisfies (1.12).

(c) Prove that if the quadratic equation $x^2 = ax + b$ has *equal* rots $\alpha = \beta$, then for A and B as above the sequence

$$A, A\alpha + B\alpha, A\alpha^2 + 2B\alpha^2, \ldots, A\alpha^n + nB\alpha^n, \ldots$$

satisfies (1.12).

We are now able to present a method for the solution of (1.12), that is for obtaining a formula for $f_n(a, b)$:

First solve the quadratic equation $x^2 = ax + b$. Then distinguish two cases.

Case 1. If the solutions α and β of the equation are different then, as in Problem 22(b), express $f_n(a, b)$ in the form

$$f_n(a, b) = A\alpha^n + B\beta^n \tag{1.13}$$

where A and B are numbers which must be determined. This can be done with the help of the given initial terms $f_0(a, b)$ and $f_1(a, b)$ of the sequence $F_n(a, b)$, as follows.

Since (1.13) is valid for all $n \geqslant 0$, rewrite it for the particular cases $n = 0$ and $n = 1$, that is

$$\left. \begin{array}{l} f_0(a, b) = A + B \\ f_1(a, b) = \alpha A + \beta B \end{array} \right\} \tag{1.14}$$

A and B are calculated by solving the system of simultaneous equations (1.14). This system has a unique pair of solutions A, B. (Why?)

Hence $f_n(a, b)$ is uniquely determined in terms of A, B, and n.

Case 2. If the solutions α and β of the equation $x^2 = ax + b$ are equal then, as in Problem 22(c), express $f_n(a, b)$ in the form

$$f_n(a, b) = A\alpha^n + nB\alpha^n \qquad (1.15)$$

and find A, B by rewriting (1.15) in the special cases $n = 0$ and $n = 1$.

We can now practise the acquired method to solve some of the recursion formulae met earlier in this chapter.

Problem 23

(a) Deduce Binet's formula, (1.6), by solving the recurrence relation $f_n = f_{n-1} + f_{n-2}$ for the Fibonacci numbers.

(b) Solve the recurrence relations for $\bar{d}_n$, $\bar{d}_n'$, $\bar{s}_n$, and $\bar{s}_n'$, obtained as solutions of Problem 16(a)–(d).

II Patterns of dots and partitions of integers

Introduction

Since antiquity patterns of dots have played an important role in the theory of numbers. The followers of Pythagoras (around 540 BC) represented certain integers by sets of dots arranged in the shape of polygons or polyhedra; such integers are nowadays called *figurate numbers*.

Examples of figurate numbers are

the *triangular numbers* t_n 1,3,6,10,...,
the *square numbers* s_n 1,4,9,16,..., and
the *pentagonal numbers* p_n 1,5,12,22,...

represented by triangular, square, respectively pentagonal arrays of dots for $n = 1,2,3,...$ (see Fig. 1.11).

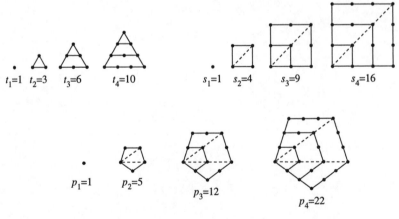

Fig. 1.11

The geometrical representation of figurate numbers affords a quick insight into the structure of these integers. Here are a few examples:

(i) By placing next to one another two triangular arrays — one 'upside down' — representing the same triangular number t_n, we obtain a

20

parallelogram of $n(n+1)$ dots (Fig. 1.12). Thus one immediately deduces the formula for t_n:

$$t_n = \frac{1}{2}n(n+1).$$

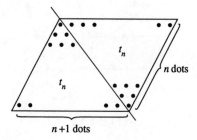

Fig. 1.12

(ii) Figure 1.13 shows a partition of the array, representing the square number s_n into corridors with $1, 3, 5, \ldots, 2n-1$ dots. This leads to the representation of any square number n^2 as the sum of the first n odd natural numbers:

$$n^2 = 1 + 3 + 5 + \ldots + (2n-1).$$

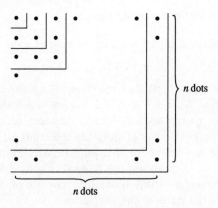

Fig. 1.13

(iii) By cutting off from the pentagonal array corresponding to p_n, for $n \geqslant 2$, a triangular array representing t_{n-1} (Fig. 1.14), one realizes that

$$p_n = t_{n-1} + s_n.$$

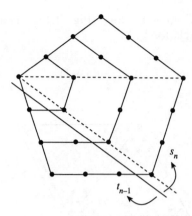

Fig. 1.14

Hence

$$p_n = \frac{(n-1)n}{2} + n^2,$$

that is

$$p_n = \frac{3n^2 - n}{2},$$

The last two examples are related to *partitions of natural numbers,* an important topic in modern number theory:

If a natural number n is expressed as the sum of some natural numbers,

$$n = n_1 + n_2 + \ldots + n_k$$

then we say that n is partitioned into parts $n_1, n_2, \ldots, n_k$.

For example, $6 = 1 + 1 + 4$ or $6 = 2 + 4$ are two distinct partitions of 6.

As in the times of the ancient Greeks, modern mathematicians find it very helpful to apply patterns of dots for the study of number partitions. The patterns used nowadays for this purpose are known as *Ferrer graphs* (or Young tableaux, in quantum mechanics). Figure 1.15 shows the Ferrer graphs of the above partitions of 6; they consist of 6 dots, arranged in rows, representing the parts of the partitions.

$$
\begin{array}{ll}
1 \to \bullet & \qquad 2 \to \bullet \ \bullet \\
1 \to \bullet & \qquad 4 \to \bullet \ \bullet \ \bullet \ \bullet \\
4 \to \bullet \ \bullet \ \bullet \ \bullet & \\
\quad\ 1+1+4 = 6 & \qquad\quad 2+4 = 6
\end{array}
$$

Fig. 1.15

Our aim in this chapter is to solve a number of problems, some of them related to the ancient patterns of figurate numbers, and some to the modern Ferrer graphs.

In Section 2.1 we study puzzles and exercises involving figurate (mostly triangular) numbers. In Section 2.2 we study Ferrer graphs of those partitions of natural numbers in which the parts all differ from one another. Our findings in Section 2,2 are used in Section 2.3 for the proof of the following famous formula of Leonhard Euler (1707–83), one of the greatest mathematicians of all time:

$$\prod_{n=1}^{\infty} (1 - x^n) = 1 + \sum_{i=1}^{\infty} (-1)^i \left(x^{(3i^2 - i)/2} + x^{(3i^2 + i)/2} \right).$$

(The symbol $\prod_{n=1}^{\infty} (1 - x^n)$ stands for the infinite product

$$(1 - x)(1 - x^2)(1 - x^3)\ldots).$$

2.1 A puzzle on a heap of pebbles and other questions

Let us start with the following simple puzzle:

Problem 24

Divide a heap of 11 pebbles into two non-empty heaps in an arbitrary way. Multiply the numbers of the pebbles in the two heaps and record the product, say π_1. Divide each of the two new heaps into two non-empty heaps (if possible), multiply the numbers of the pebbles in the two parts of the divided heaps, and again record the products, π_2 and π_3 (Fig. 1.16). Continue this procedure: divide each newly obtained heap into two non-empty heaps (if possible) and record the product of the numbers of pebbles in the two 'sub-heaps'. At the end 11 heaps will be obtained, with a single pebble in each. Here the process stops. At this stage calculate the sum s of all products:

$$s = \pi_1 + \pi_2 + \pi_3 + \ldots.$$

How should one divide the heaps into two new heaps at each stage so that the resulting sum s of the products attains its maximal value?

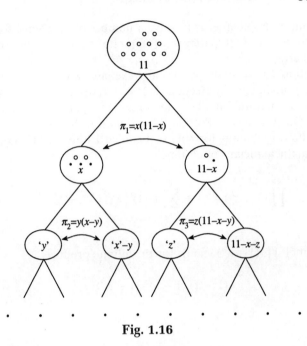

Fig. 1.16

Problem 24 is not very difficult. Having solved it try to find a geometrical interpretation of the solution. This may help to explain now only why your solution is correct but also why it was to be expected.

Problem 25

(a) Give a geometrical explanation of the solution of Problem 24.

(b) Generalize Problem 24 for a heap of n pebbles where n is any integer greater than 1.

The next two problems are variants of Problem 25.

Problem 26

A heap of $n \geqslant 3$ pebbles is divided into three non-empty heaps with say x, y and $z = n - (x + y)$ pebbles respectively. Form the expression $\varepsilon_1 = xy + xz + yz$. Continue this procedure; at each stage divide the newly obtained heaps into three non-empty heaps (if possible) and record the corresponding expression ε_j. If one of the heaps contains only two pebbles, split it into two heaps with one pebble each; the corresponding expression ε_j will be equal to $1 \cdot 1 + 1 \cdot 0 + 1 \cdot 0 = 1$.

At the end of the procedure, when n heaps with a single pebble each are obtained, calculate the sum

$$s = \varepsilon_1 + \varepsilon_2 + \dots.$$

Does the value s depend on the way the heaps are divided into smaller ones? Interpret your result geometrically using arrays of dots.

Problem 27

Divide a heap of $n \geqslant 2$ pebbles into two non-empty heaps with x and y pebbles respectively, and form the expression $\zeta_1 = x^2 y + xy^2$. Continue this procedure with each new heap until n heaps with one pebble each are obtained. Does the sum $s = \zeta_1 + \zeta_2 + \dots$ depend on the way the heaps are divided?

Try to interpret your result geometrically using three-dimensional shapes. (*Hint.* It may be more appropriate to use solid shapes — cuboids and pyramids with edge lengths $n, x, y, \dots$ instead of patterns of dots.)

It is quite natural to try to generalize a given problem or formula in order to understand better its significance and effectiveness. One should not, however, ignore the possible benefits of the study of special cases. This is illustrated by the next two problems. By applying Problems 25–7 to heaps in which the numbers of the pebbles are Fibonacci and 3-bonacci numbers we shall obtain new sets of identities for these numbers.

Problem 28

In Problems 25 and 27 take for n a Fibonacci number, say $f_k \geqslant f_2$. Divide the heap of f_k pebbles into two heaps so that one of them contains f_{k-1} and the other f_{k-2} pebbles. (This can be done since $f_k = f_{k-1} + f_{k-2}$ for $k \geqslant 2$.) After that divide the heap of f_{k-2} pebbles into two heaps with f_{k-3} and f_{k-4} pebbles respectively, and the heap of f_{k-1} pebbles into two heaps with f_{k-2} and f_{k-3} pebbles (if possible). Continue this process — at each stage dividing a heap of f_i pebbles into two heaps with f_{i-1} and f_{i-2} pebbles — until all heaps contain only one pebble.

Rewrite the formulae obtained for s in Problems 25 and 27 for these special numbers of pebbles in the heaps. This will yield two identities for Fibonacci numbers.

Problem 29

In Problem 26 take for n a 3-bonacci number — say $f_k(3) \geqslant f_3(3)$. Divide the heap of $f_k(3)$ pebbles into three heaps so that they contain $f_{k-1}(3), f_{k-2}(3)$, and $f_{k-3}(3)$ pebbles respectively. At each successive stage divide a heap containing $f_i(3)$ pebbles into three heaps with $f_{i-1}(3), f_{i-2}(3,$ and $f_{i-3}(3)$ pebbles respectively (until all heaps contain only one pebble).

Rewrite the formula for s obtained in Problem 26 for these numbers of pebbles in the heaps. This will yield a relation for 3-bonacci numbers.

Figurate numbers appear in different fields of mathematics, often rather unexpectedly. The last problem in this section describes a connection between triangular numbers and the structure of regular $2n$-gons.

Problem 30

Figure 1.17 shows the partition of a regular hexagon into three rhombuses, and of a regular octagon into six rhombuses.

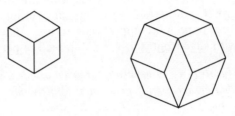

Fig. 1.17

(a) Prove that, for any integer $n \geqslant 2$, a regular $2n$-gon can be divided into t_{n-1} rhombuses where t_{n-1} is the $(n-1)$st triangular number.

(b) Calculate the angles of these rhombuses in terms of n.

(c) Using (a) and (b) prove the trigonometric formula

$$\sum_{i=1}^{n-1} (n-i) \sin \frac{i \cdot 180°}{n} = \frac{n}{2} \cotan \frac{90°}{n}.$$

2.2   Partitions of integers into distinct parts and corresponding Ferrer graphs

The representation of a natural number n as a sum of a certain number of natural numbers $p_1, p_2, \ldots, p_k$:

$$n = p_1 + p_2 + \ldots + p_k \qquad (1.16)$$

is called a *partition* of n; the summands $p_1, \ldots, p_k$ are the *parts* of this partition.

The order in which the parts follow one another in a partition is irrelevant; therefore we shall arrange them so that no part is followed by a smaller one. That is:

$$1 \leqslant p_1 \leqslant p_2 \leqslant \ldots \ldots \leqslant p_k \leqslant n.$$

The *Ferrer graph* of the partition (1.16) consists of n evenly spaced dots arranged in k rows and p_k columns. The first row consists of p_1 dots, the second of p_2 dots — with its first dot underneath the first dot of the first row — and in general the ith row consists of p_i dots, with its first dot underneath the first dot of the previous row. For example, the Ferrer graph of the partition $12 = 2 + 2 + 3 + 5$ is given in Fig. 1.18.

Fig. 1.18

The great advantage of Ferrer graphs for studying partitions is the following: by a suitable transformation of the pattern of dots, the Ferrer graph of a partition of n can be changed into a Ferrer graph representing a different partition of the same natural number n. Thus one can compare properties of various partitions of n by studying their Ferrer graphs. Here is an example.

The Ferrer graph of the partition $12 = 1 + 1 + 2 + 4 + 4$ is transformed into the Ferrer graph of the partition $12 = 2 + 2 + 3 + 5$ as shown in Fig. 1.19.

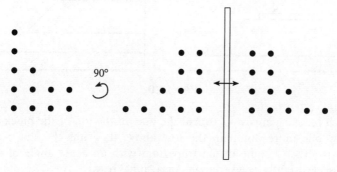

Fig. 1.19

In Fig. 1.19 the Ferrer graph of the partition $p_1 + p_2 + \ldots + p_k$ of n has been transformed (by a rotation through $90°$ followed by a reflection in a vertical axis) into another Ferrer graph corresponding to a different partition $p_1' + p_2' + \ldots + k$ of n. The important feature of this transformation is that it changes the k rows of the original graph into k columns of the resulting graph.

It is now easy to prove the following interesting statements.

Problem 31

Let n be any natural number and let k be any natural number not greater than n. Prove that:

(a) The number of partitions of n into k parts is equal to the number of partitions of n into integers the largest of which is k.

(b) The number of partitions of n into at most k parts is equal to the number of partitions of n into parts which do not exceed k.

We shall now turn to the study of partitions of n in which all parts are different. Thus in what follows $P(n; p_1, p_2, \ldots, p_k)$ will denote the partition

$$n = p_1 + p_2 + \ldots + P_k \quad \text{with} \quad 1 \leqslant p_1 < p_2 < \ldots < p_k \leqslant n .$$

The corresponding Ferrer graph will be denoted by $F(n; p_1, p_2, \ldots, p_k)$.

Since all parts p_i are different, no two rows of the corresponding Ferrer graph are equal. The graph can be divided into one or more blocks as shown in Fig. 1.20.

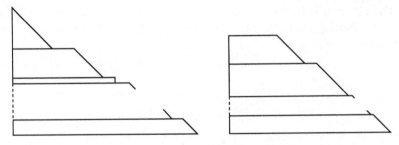

Fig. 1.20

In each block each row — except the one on the top of the block — has exactly one more dot than the row above it. (Thus the shape of any block is either a right-angled trapezium with an acute angle of 45°, or an isosceles right-angled triangle, or a single row.)

For any two consecutive blocks the first row of the lower block contains at least 2 more dots than the last row of the upper block.

Our next aim is to use the Ferrer graphs $F(n; p_1, p_2, \ldots, p_k)$ for comparing the number of *'odd partitions'* of n (that is partitions into an odd number of parts) with the number of the *'even partitions'* of n (partitions into an even number of parts).

This will be done with the aid of two transformations, α and β, defined as follows.

α *is the transformation that shifts the p_1 dots of the top row of a Ferrer graph behind the last dots of the p_1 bottom rows — one into each of the bottom p_1 rows (see Fig. 1.21(a)).*

β *is the transformation that shifts the last dot from each of the rows of the lowest block of a Ferrer graph into a single row above its top row (Fig. 1.21(b)).*

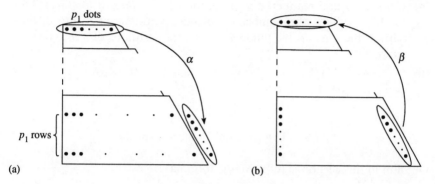

Fig. 1.21

It is important to note that:

(a) The image of a Ferrer graph $F(n; p_1, p_2, \ldots, p_k)$ is not necessarily a Ferrer graph; however,

(b) if α transforms $F(n; p_1, p_2, \ldots, p_k)$ into another Ferrer graph $F(n; p_1', p_2', \ldots, p_{k'}')$ then it changes the parity in the number of its rows; that is, if k is even, then k' will be odd and vice versa, if k is odd then k' will be even. The same holds for β.

We shall say that a Ferrer graph $F(n; p_1, p_2, \ldots, p_k)$ *admits* α or β if the image $\alpha(F)$, respectively $\beta(F)$, is also a Ferrer graph $F(n; p_1', p_2', \ldots, p_{k'}')$.

Now we are able to solve the following problems, crucial for the proof of Euler's formula in the next section.

Problem 32

Prove that:

(a) If n is a natural number such that n is *not* equal to $(3i^2 - i)/2$ or $(3i^2 + i)/2$ for any natural number i, then each partition of n into distinct parts has a Ferrer graph admitting either α or β. (*Hint.* Consider separately Ferrer graphs consisting of more than one block, and Ferrer graphs with one block only.)

(b) If $n = (3i^2 - i)/2$ or $n = (3i^2 + i)/2$ for some natural number i then exactly one partition of n into distinct parts admits neither α nor β.

Problem 33

Applying the solutions of Problem 32(a) and (b), deduce the following.

(a) If n is a natural number, not of the form $(3i^2 - i)/2$ or $(3i^2 + i)/2$ for $i = 1, 2, \ldots$, then the number O_n of odd partitions of n into distinct parts is the same as the number E_n of even partitions of n into distinct parts.

(b) If $n = (3i^2 - i)/2$ or $n = (3i^2 + i)/2$, then $E_n - O_n$ is equal to $+1$ if i is even, and to -1 if i is odd.

2.3   The proof of Euler's formula

Our aim in this section is to prove the formula

$$\prod_{n=1}^{\infty}(1 - x^n) = 1 + \sum_{i=1}^{\infty}(-1)^i\left(x^{(3i^2-i)/2} + x^{(3i^2+i)/2}\right). \qquad (1.17)$$

To get an idea for the proof one should study the expansions of the partial products

$$\Pi_n = (1 - x)(1 - x^2)\ldots(1 - x^n).$$

Problem 34

Expand Π_n by multiplying the factors and collecting like terms. Prove that in this expansion the coefficient of x^m for $m \leqslant n$ is equal to $O_m - E_m$ (for the definition of O_m and E_m see Problem 33).

It is now easy to complete the proof of Euler's theorem:

Problem 35

Deduce from the solutions of Problems 33 and 34 the validity of formula (1.17).

Remark. Note the following curious fact. The exponents $(3i^2 - i)/2$ of x in Euler's formula are the pentagonal numbers $1, 5, 12, 22, \ldots$.
The remaining exponents in the formula form the sequence $2, 7, 15, 26, \ldots$.
The numbers in the set $\{1, 5, 12, 22, \ldots\} \cup \{2, 7, 15, 26, \ldots\}$ are often called the *generalized pentagonal numbers*.

III Patterns, related to rational numbers; periodic decimal fractions, repunits, and visible lattice points

Introduction

Numbers expressible as fractions a/b, where a and b are integers and $b > 0$, are called *rational numbers*.

Each rational number can be expanded as a decimal fraction (e.g. $2/5 = 0.4, 5/3 = 1.666\ldots$) with either a finite or infinite number of decimal places. In Sections 3.1 and 3.2 we shall study sequences of digits in decimal expansions of rational numbers; we shall also deduce a divisibility criterion for *repunits* — important numbers in computer science and coding theory.

Section 3.3 discusses the representation of fractions by lattice points of a two-dimensional cartesian coordinate system. The *lattice points* of the coordinate system are the points with integer coordinates. The lattice point with coordinates (b, a), where $b \neq 0$,, is assigned to the fraction a/b. We shall look at patterns of lattice points assigned to reduced fractions.

3.1 Periodic decimal numbers; what is a repunit?

In this section, a/b will denote a *reduced fraction*, that is a fraction in which a and b have no common factor greater than 1. Moreover, a/b will be positive and less than 1.

It is easy to find out which fractions a/b have in their decimal expansions a finite or infinite number of non-zero digits:

Problem 36

The fraction a/b has an infinite number of digits in its decimal expansion if and only if the denominator b has at least one prime factor different from 2 and 5.

Prove this statement.

31

Decimal fractions with infinitely many digits are called *infinite* decimal fractions. Here are two examples:

$$\tfrac{1}{7} = 0.142857142857\ldots, \text{ written briefly as } 0.\dot{1}4285\dot{7}, \text{ and}$$

$$\tfrac{1}{6} = 0.1666\ldots \qquad\qquad, \text{ written briefly as } 0.1\dot{6}.$$

In both examples, from a certain point on, a group of digits keeps on repeating itself. Such infinite decimal fractions are called *periodic*. The group of repeating digits in a periodic decimal fraction is called the *period,* and the number of digits in the period is the *length of the period.* (Thus the period of $0.\dot{1}4285\dot{7}$ is 142857, of length 6; the period of $0.1\dot{6}$ is 6, of length 1.)

We shall distinguish between *purely periodic* and *mixed periodic* fractions; in the former the period starts immediately after the decimal point and in the latter there are one or more digits between the decimal point and the period. ($0.\dot{1}4285\dot{7}$ is purely periodic, while $0.1\dot{6}$ is mixed periodic.)

We ask: Is a fraction with an infinite decimal expansion necessarily periodic? If so, when is such a periodic fraction purely periodic? Answers are provided by solutions of the following two problems.

Problem 37

Prove that if a fraction a/b has an infinite decimal expansion, then this expansion is periodic, and that the period of the expansion consists of at most $b - 1$ digits.

Problem 38

Prove that if the denominator b of a fraction a/b has no factor equal to 2 or 5 then its decimal expansion is purely periodic.

The following property of purely periodic decimal fractions leads us to *repunits,* that is natural numbers whose digits are all equal to 1: 1, 11, 111, 1111, ... (The term 'repunit' was coined as an abbreviation of the expression 'repeated unit'.)

Problem 39

Prove that if a/b has a purely periodic decimal expansion, then the length of the period is the smallest number L for which $10^L - 1$ is divisible by b. *Remark.* Note that L does not depend on the numerator a of the fraction.

The number $10^L - 1$ in Problem 39 consists of L digits, all equal to 9:

$$10^L - 1 = \underbrace{99\ldots.9}_{L}$$

Thus $10^L - 1 = 9 \cdot \underbrace{11\ldots.1}_{L}$ where $\underbrace{11\ldots.1}_{L}$ is a repunit.

Numbers with digits 1 and 0 play an important role in various branches of modern mathematics, for example in coding theory and in computer science. Hence there is also an interest in repunits. The following divisibility condition for repunits can be deduced from Problem 39:

Problem 40

Prove that if b is not divisible by 2, 5, or 3 then among the multiples of b there is a repunit.

In other words:

For any natural number b coprime to 2, 5, and 3 there exists at least one repunit which is divisible by b.

It is interesting that in the eighteenth century, long before repunits had a recognized practical or theoretical significance, the celebrated Swiss mathematician Jacob Bernoulli (1654–1705) spent a considerable time in finding prime factors of repunits. He constructed a table showing factors of repunits $\underbrace{11\ldots.1}_{n}$ for most values of $n \leqslant 31$. In recent decades, thanks to computers and to renewed interest in repunits, Bernoulli's table has been greatly extended — including numbers with thousands of digits. Researchers are also looking for repunits which are prime numbers.

Problem 41

According to the statement of Problem 39, if b is coprime to 10, then the periods of all fractions a/b have the same length. But what can one say about the digits of the periods of fractions a/b with common b but different values of a?
Investigate sets of fractions such as

$$\frac{1}{7}, \frac{2}{7}, \ldots, \frac{6}{7} \text{ or } \frac{1}{13}, \frac{2}{13}, \ldots, \frac{12}{13}.$$

Can you reach any conclusion?

3.2   Exercises on decimal expansions

Problem 42

Let n be a natural number. Prove that the decimal expansion of the sum of the fractions

$$\frac{1}{n} + \frac{1}{n+1} + \frac{1}{n+2}$$

is *not* purely periodic.

Problem 43

Is the number $0.123456789101112131415\ldots$, obtained by writing successively all the non-negative integers, a periodic decimal?

Problem 44

(a) Prove that the expression

$$a_n = \frac{n^3 + 2n}{n^4 + 3n^2 + 1}$$

where n is any natural number, is a fraction in its lowest terms.

(b) For which values of n does a_n represent a periodic infinite decimal?

Problem 45   *(reference [12], 1907/3)*

Let $r/s = 0.k_1 k_2 k_3 \ldots$ be the decimal expansion of a rational number. (If this is a terminating decimal, all the k_i from a certain point on are 0.)

Prove that at least two of the numbers

$$\sigma_1 = 10\frac{r}{s} - k_1, \; \sigma_2 = 10^2 \frac{r}{s} - (10k_1 + k_2),$$

$$\sigma_3 = 10^3 \frac{r}{s} - (10^2 k_1 + 10k_2 + k_3), \ldots$$

are equal.

3.3   Fractions and lattice points

Consider a two-dimensional cartesian coordinate system. The points with integer coordinates in this system are the points of a square grid called a square lattice. The vertices of this lattice — that is, the points with integer coordinates — are the lattice points (Fig. 1.22).

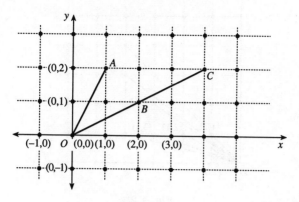

Fig.1.22

With each fraction m/n we shall associate the lattice point (n, m). Our aim is to discover properties of fractions by studying sets of lattice points associated with fractions.

A lattice point P is called *visible* (from the origin $O(0, 0)$) if there is no lattice point between O and P on the straight line segment OP.

(In Fig. 1.22 the points A and B are visible lattice points and C is not.)

It is easy to show the following:

Problem 46

A lattice point associated with a fraction m/n is visible if and only if m/n is a reduced fraction, that is, if and only if m and n are coprime.

The fractions associated with the visible points on or inside the triangles with vertices $O(0, 0)$, $A_n(n, 0)$, and $B_n(n, n)$ form intriguing sequences, known as Farey series (F_n).

Imagine that a light ray, emerging from O and initially making an angle of $0°$ with the positive direction of the x-axis, sweeps out the angle A_4OB_4. The ray hits the visible lattice points of triangle A_4OB_4 (starting with $(1, 0)$ and ending with the point $(1, 1)$ on OB_4) (see Fig. 1.23). The corresponding fractions form the 'fourth Farey series' F_4:

$$\frac{0}{1}, \frac{1}{4}, \frac{1}{3}, \frac{1}{2}, \frac{2}{3}, \frac{3}{4}, \frac{1}{1}.$$

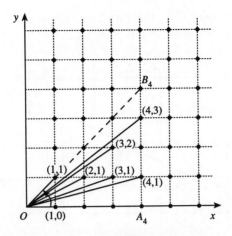

Fig. 1.23

Problem 47

Write down the Farey series F_1, F_2, F_3, F_4, F_5, and F_6 in six rows underneath one another and investigate the properties of this number triangle.

Can you generalize some of your observations?

Remark. The following property of parallelograms, with lattice points as vertices, can be helpful in the study of the Farey series:

ℙ Let P be a parallelogram whose vertices are lattice points. If there are no other lattice points on the perimeter of P or in its interior, then the area of P is 1.

ℙ is a special case of a very useful statement on lattice polygons known as *Pick's theorem:*

> *If the vertices of a polygon P are lattice points then its area is $p/2 + i - 1$, where p is the number of lattice points on the perimeter and i the number of lattice points in the interior of P.*

For the proof of Pick's theorem see for example Honsberger [11]; a proof of ℙ will be given in Chapter IX, Section 9.3.

IV Reflected light rays and real numbers; a theorem of Kronecker

Introduction

A light ray is reflected from a straight line mirror in such a way that the angle between the incoming ray and the mirror is equal to the angle between the reflected ray and the mirror; moreover, the incoming ray, the mirror, and the reflected ray belong to a common plane (Fig. 1.24).

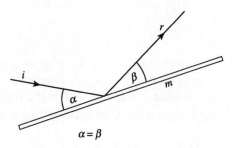

$$\alpha = \beta$$

Fig. 1.24

In Section 4.1 we shall investigate the path of a light ray after repeated reflections, when two or more straight lines or plane mirrors are used. In particular, we shall consider paths that form closed polygonal lines.

In Section 4.2 we deal with the following question: What can be said about the path of a light ray, reflected repeatedly from four mirrors that form a square, if the path is not a closed polygonal line? To answer this question we shall use a theorem on real numbers due to Kronecker (1823–91). Kronecker's theorem will be stated and its proof will be outlined in the first part of Section 4.2.

The last section, 4.3, of this chapter contains problems on real numbers with solutions deduced from Kronecker's theorem.

4.1 Problems on reflected light rays

Problem 48

m_1 and m_2 are two straight-line mirrors in a common plane, and P and Q are two points between m_1 and m_2 in the plane.

Construct the path of the light ray which emerges from P and, after reflection from m_1 and m_2, passes through Q.

(Discuss the solutions of this problem associated with different positions of the mirrors and of the points.)

Problem 49

m_1 and m_2 are two parallel straight-line mirrors; P and Q are two points between m_1 and m_2.

(a) Construct the path of the light ray which emerges from P, is reflected four times (twice from m_1 and twice from m_2), and then passes through Q.

(b) Let n be an arbitrary natural number. Can a light ray, emerging from P pass through Q after a total of n reflections from m_1 and m_2? If so, describe a method for the construction of the path.

Problem 50

(a) m_1 and m_2 are two straight-line mirrors forming an angle α. A light ray approaches m_1 in a given direction d and is reflected once from each mirror (Fig. 1.25). For what values of α is its final direction equal to $-d$?

(b) Answer the question in (a) if the light ray is reflected from each mirror twice.

(c) Generalize this problem.

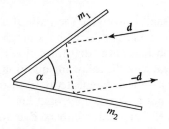

Fig. 1.25

Problem 51

(a) The sides of a square $ABCD$ are mirrors; L is a light source on the side DC. Can a light ray emerging from L hit all four sides of the square, return to L, and continue to travel repeatedly along the same path?

(b) The faces of a cube $ABCDEFGH$ are mirrors. A light source L is placed on the face $ABCD$. Can a light ray emerging from L hit all six faces of the cube, return to L and continue to travel along the same path?

Problem 52

The sides of a square *ABCD* are mirrors; *L* is a light source on *AB* and a light ray emerges from *L* at an angle α to *AB* (Fig. 1.26).

Prove that the path of this light ray will form a closed polygonal line *P* (which can be self-intersecting) if and only if tan α is a rational number.. Show that in this case the path is periodic, that is the light ray travels invariably along *P*.

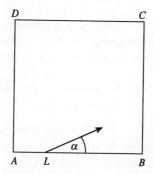

Fig. 1.26

Hint. The path of any light ray inside the square *S = ABCD* can be mapped onto a straight line as shown in the example below:

Suppose that the light ray, emerging from *L*, hits *BC* at *P*, *AD* at *Q*, *DC* at *R*, and *BC* at *T*. (Fig. 1.27). Construct the reflection of *S* in *CB* to obtain $S_1 = A_1BCD_1$, then the reflection of S_1 in A_1D_1 to obtain $S_2 = A_1B_2C_2D_1$, followed by the reflection of S_2 in D_1C_2 to obtain $S_3 = A_3B_3C_2D_1$, and by the reflection of S_3 in B_3C_2 to obtain

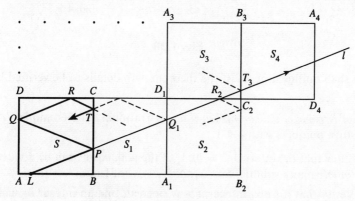

Fig. 1.27

$S_4 = A_4B_3C_2D_4$. The points L and P and the images $Q_1 \in S_1$ of Q, $R_2 \in S_2$ of R and $T_3 \in S_3$ of T lie on a common straight line ℓ. Study the position of ℓ in the plane when the path of the light ray is a closed polygonal line.

4.2   Non-periodic light ray paths; a theorem of Kronecker on irrational numbers

The statement in Problem 52 raises the question: How does a light ray travel inside a square S, bounded by four mirrors, if its path is not a closed polygonal line (that is, if $\tan\alpha$, for α defined in Problem 52, is irrational)?

The above question will be answered with the help of the following theorem of Kronecker on irrational numbers.

The theorem of Kronecker: If ϑ is an irrational number then the set of all numbers $n\vartheta + m$, where n and m are integers, and n is greater than a given positive number N, is dense on the real axis $\mathbb{R}$.

'Dense on the real axis' means that every neighbourhood of any point of $\mathbb{R}$ contains a point with coordinate $n\vartheta + m$; that is

If $r \in \mathbb{R}$ and ε is a given positive number, then there exist integers m and n, where $n > N$, such that

$$r - \frac{\varepsilon}{2} < m\vartheta + n < r + \frac{\varepsilon}{2}.$$

(See Fig. 1.28.)

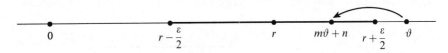

Fig. 1.28

We shall outline a proof of the theorem with details to be verified by the reader.

Proof of Kronecker's theorem. Let ϑ be an irrational number and let N and ε be positive numbers with $\varepsilon < 1$.

(i)   Show that the interval $\Delta = [0, 1]$ of the real axis $\mathbb{R}$ can be divided into *finitely* many subintervals $\Delta_1, \Delta_2, \ldots, \Delta_k$ of length less than ε.

(ii) Verify that for any integer $n > N$ one can find an integer m such that $n\vartheta + m \in \Delta$.

(iii) Show that there are *infinitely* many numbers $a_1 = n_1 \vartheta + m_1$, $a_2 = n_2 \vartheta + m_2, \ldots$ in Δ such that m_i are integers and n_i are integers with $n_1 > N$ and $n_{i+1} - n_i > N$.

(iv)   Using (iii) show that at least one of the subintervals of Δ, say Δ_ℓ, must contain two of the numbers $a_1, a_2, \ldots$, say a_i and a_j.

(v)   Show that if in (iv) $j > i$, then the difference $a' = a_j - a_i$ is a non-zero number of the form $n' \vartheta + m'$ and n' are integers and $n' > N$. Moreover $|a'| < \varepsilon$.

(vi)   Consider the set of points on the real axis $\mathbb{R}$ with coordinates $0, a', 2a', 3a', \ldots$. If $a' > 0$ then these points divide the non-negative part of $\mathbb{R}$ into intervals of length less than ε; if $a' < 0$ then $0, a', 2a', 3a', \ldots$ divide the non-positive part of $\mathbb{R}$ into such small intervals.

In the first case, by adding negative integers to $0, a', 2a', 3a', \ldots$, the intervals between them are translated into the negative part of the real axis; in the second case, by adding positive integers to $0, a', 2a', 3a', \ldots$, the intervals are translated into the positive part of $\mathbb{R}$.

Thus the whole real axis $\mathbb{R}$ is covered by intervals of length $|a'| < \varepsilon$. It follows readily that Kronecker's theorem holds.

Remark. It is instructive to prove that the requirement that ϑ should be irrational is essential for Kronecker's theorem to hold:

Problem 53

Show, by giving an example, that if in Kronecker's theorem ϑ is replaced by a rational number, then the statement of the theorem is no longer true.

We are now ready to investigate the path of a light ray inside a square bounded by mirrors in the case not covered by Problem 52.

Problem 54

Let $S = ABCD$ be a square bounded by mirrors, and let L be a light source on AB. Prove that if a light ray emerges from L, making an angle α with AB, where $\tan \alpha$ is irrational, then its path runs arbitrarily near to any point of the square.

Hint. Set up a cartesian coordinate system in which the vertices of the square S have coordinates $(0,0)$, $(1,0)$, $(1,1)$, and $(0,1)$ (Fig. 1.29) and let the coordinates of the light source L be a and 0.

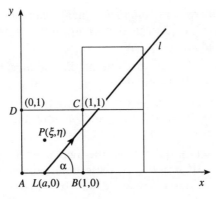

Fig. 1.29

As in the Hint for Problem 52 (p. 39), map the points of the path of the light ray onto points of the line ℓ through L with gradient $\tan \alpha$. The line ℓ passes through the images $s_1, s_2, s_3, \ldots$ of $S, S_1, S_1, \ldots$ respectively, obtained by reflections in those of their sides which are met by L.

Let $P(\xi, \eta)$ be any point of S. Prove that there exists an image $P' \in S_i$ of P with coordinates of the form $\xi + 2m$, $\eta + 2n$ for appropriate integers m and n such that an arbitrarily small neighbourhood of P_i contains a point of L with the same y-coordinate $\eta + 2n$. This can be achieved by applying the theorem of Kronecker.

4.3   Further applications of Kronecker's theorem

Problem 55

In the first five numbers of the sequence $11^0, 11^1, 11^2, \ldots$ the digits form a nice pattern related to Pascal's triangle:

$$
\begin{aligned}
11^0 &= \quad 1 \\
11^1 &= \quad 1\ 1 \\
11^2 &= \quad 1\ 2\ 1 \\
11^3 &= \ 1\ 3\ 3\ 1 \\
11^4 &= 1\ 4\ 6\ 4\ 1
\end{aligned}
$$

Unfortunately, here the pattern breaks down: $11^5 \neq 15101051$. Nevertheless, one is tempted to ask: Is there a power of 11 starting with the digits 15101051?

Problem 56

There are restrictions on the last digit of a square number: it cannot be 2, 3, 7, or 8. Are there any restrictions on the starting digits of square numbers?

Find out: If $a_1a_2a_3a_4\ldots a_s$ is a sequence of digits such that $a_1 \neq 0$ and $a_i \in \{0,1,2,\ldots,9\}$ for $i = 2,\ldots,s$, is there a square number whose first s digits are $a_1a_2\ldots a_s$?

Problem 57

Consider the trigonometric function

$$f(x) = \sin \pi x + \sin \pi\sqrt{2}x$$

defined on the set of real numbers.

Prove that

(a) $f(x)$ cannot reach the value 2 for any real number x; however,

(b) there are real numbers x for which $f(x)$ is arbitrarily close to 2.

Remark. The function in Problem 57 is continuous at every real number r, that is

$f(r)$ is defined for r (it has a unique value), and for every positive number ε (however small) a positive number δ (depending on ε) can be found such that

$$-\varepsilon < f(x) - f(r) < \varepsilon$$

for all real numbers x which satisfy the relations

$$-\delta < x - r < \delta.$$

(see Fig. 1.30)

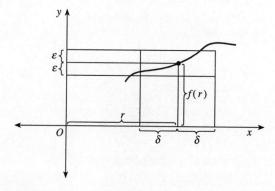

Fig. 1.30

The following statement about continuous functions has many important applications in mathematics.

(*) *Let S be a dense subset of the set $\mathbb{R}$ of the real numbers. If two functions $f(x)$ and $g(x)$, continuous for all $r \in \mathbb{R}$, take equal values on S (that is $f(a) = g(a)$ for every $a \in S$), then $f(x) = g(x)$ for all real numbers x.*

Using (*) and Kronecker's theorem, we are able to solve the following problem:

Problem 58

Let ϑ be a given irrational number. Find all continuous real functions $f(x)$ such that

$$f(x) = f(x + 1) = f(x + \vartheta) \quad \text{for all} \quad x \in \mathbb{R}.$$

V The Chinese remainder theorem and invisible lattice points

Introduction

The fourth-century Chinese mathematical text, titled *Sun Tsu San Ching (Master Sun's arithmetic manual)*, contains the following problem:

> There is an unknown number of objects. When counted in 'threes', the remainder is 2; when counted in 'fives', the remainder is 3; and when counted in 'sevens', the remainder is 2. How many objects are there?

Sun Tsu describes a way for solving the problem and provides an answer to the question. The number of objects is 23.

In modern notation, if x is the number of objects, the conditions of Sun Tsu's problem can be formulated as three indeterminate linear equations:

$$\left.\begin{array}{l} x = 3q_1 + 2 \\ x = 5q_2 + 3 \\ x = 7q_3 + 2 \end{array}\right\} \qquad (1.18)$$

where q_1, q_2, and q_3 are integers to be found. It is left to the reader to show that, apart from $x = 23$, the system has many more solutions.

Generalizations of Sun Tsu's problem were studied later by various Chinese mathematicians, culminating in the statement now known as *Chinese remainder theorem*. In modern number theory, the Chinese remainder theorem is expressed in terms of *linear congruences*. In Section 5.1 we shall introduce the notion of a linear congruence and, after proving a number of basic properties of linear congruences, we shall formulate and prove the Chinese remainder theorem.

Section 5.2 is devoted to some applications of the Chinese remainder theorem — for solving problems involving numbers and 'invisible' lattice points.

5.1   Linear congruences and the Chinese remainder theorem

Let m be a natural number and let a and b be two integers. We say that a and b are *congruent modulo m* if they yield equal remainders when divided by m (e.g. 12 and 33 are congruent modulo 7 because $12 = 1 \cdot 7 + 5$ and $33 = 4 \cdot 7 + 5$). a congruent b modulo m is written

$$a \equiv b \pmod{m};$$

the above expression is called a *congruence*. Congruences have a number of simple but very useful properties; some of them are described in the following problems.

Problem 59

Prove that:

(a) $a \equiv b \pmod{m}$ if and only if m divides the differences $a - b$.
(b) If $a_1, a_2, \ldots, a_n$ and $b_1, b_2, \ldots, b_n$ are given integers such that $a_i \equiv b_i \pmod{m}$ for a natural number m, then

$$a_1 + a_2 + \ldots + a_n \equiv (b_1 + b_2 + \ldots + b_n) \pmod{m}$$

and

$$a_1 \, a_2 \cdot \ldots \cdot a_n \equiv b_1 \, b_2 \cdot \ldots \cdot b_n \pmod{m}.$$

Problem 60

Prove that the relation of congruence mod m is

(a) *reflexive*, that is $a \equiv a \pmod{m}$ for all integers a,
(b) *symmetric*, that is if $a \equiv b \pmod{m}$ then $b \equiv a \pmod{m}$ for all integers a, b, and
(c) *transitive*, that is if $a \equiv b \pmod{m}$ and $b \equiv c \pmod{m}$ then $a \equiv c \pmod{m}$ for all integers, a, b, c.

Remark. Relations relating any pair x, y of elements of a set are called *binary relations on the set*. Binary relations which are reflexive, symmetric, and transitive are called *equivalent relations*; thus in view of Problem 60 congruence modulo m is an equivalence relation on the set of $\mathbb{Z}$ of all integers.

For $i \in \{0, 1, \ldots, m - 1\}$ denote by $\bar{i}$ the set of all integers congruent to i modulo m. The set $\bar{i}$ is called a *residue class modulo m*. Since the remainder of any integer, when divided by m, is one of the numbers $0, 1, 2, \ldots, m - 1$, it follows that every integer belongs to exactly one of the residue classes modulo m.

We are now ready to solve linear congruences and systems of linear congruences.

Let m be a given natural number, and let a and b be given integers. The relation

$$ax \equiv b \,(\text{mod } m)$$

where x is an unknown integer, is called a *linear congruence*; any integer x satisfying this congruence is called a *solution* of the congruence.

The following problem discusses the existence of solutions of linear congruences, depending on a, b, and m.

Problem 61

Prove that:

(a) If a and m are coprime, then the linear congruence

$$ax \equiv b \,(\text{mod } m) \qquad (1.19)$$

has exactly one solution modulo m.
(In other words, the solutions of (1.19) are the elements of exactly one residue class modulo m.)

(b) If a and m are not coprime, and their greatest common factor d does not divide b, then the congruence (1.19) has no solution.

(c) If a and m are not coprime, and their greatest common factor d divides b, then (1.19) has exactly d solutions modulo m (that is, the solutions of (1.19) are the elements of d residue classes modulo m).

This brings us to the Chinese remainder theorem on systems of linear congruences.

Problem 62

Prove the so-called Chinese remainder theorem:

If $m_1, m_2, \ldots, m_n$ are positive integers, that are pairwise coprime (that is, the highest common factor of m_i, m_j is 1 for any $i \neq j$, $i, j = 1, 2, \ldots, n$), and if $b_1, b_2, \ldots, b_n$ are arbitrary integers, then the system of congruences

$$\left.\begin{array}{l} x \equiv b_1 (\text{mod } m_1) \\ x \equiv b_2 (\text{mod } m_2) \\ \quad\ldots\ \ldots \\ x \equiv b_n (\text{mod } m_n) \end{array}\right\} \qquad (1.20)$$

has exactly one solution modulo the product $m_1 m_2 \cdot \ldots \cdot m_n$.

Hint. Let $M = m_1 m_2 \cdot \ldots \cdot m_n$ and $M_i = M/m_i$ for $i = 1, 2, \ldots, n$. Denote by M'_i the solution of the congruence $M_i x \equiv 1 \pmod{m_i}$ for $i = 1, 2, \ldots, n$. (Explain why M'_i exists and is unique.)

Show that

$$x = b_1 M_1 M'_1 + b_2 M_2 M'_2 + \ldots + b_n M_n M'_n \tag{1.21}$$

is a solution of each congruence of the system (1.20). Thus (1.21) is a solution of the system. Show that (1.21) is the unique solution of (1.20), modulo $m_1 m_2 \cdot \ldots \cdot m_n$.

5.2   Applications of the Chinese remainder theorem

A simple application of the Chinese remainder theorem is to verify that $x = 23$ is among the solutions of the system (1.18) corresponding to Sun Tsu's problem. This is left to the reader.

The next application, for solving a problem on coprime numbers, is more elaborate.

Problem 63   (Kvant, M1014)

(a) Let $a_1, a_2, \ldots, a_n$ be n distinct, pairwise coprime natural numbers. Prove that there exist infinitely many natural numbers b such that $b + a_1, b + a_2, \ldots, b + a_n$ are also pairwise coprime.

(b) Let $a_1, a_2, \ldots, a_n$ be n distinct natural numbers. For any prime p denote by $r_i(p)$ the remainder of a_i, $i = 1, 2, \ldots, n$, when divided by p. Prove that there exist infinitely many integers b such that $a_1 + b, a_2 + b, \ldots, a_n + b$ are pairwise coprime, if the following condition is satisfied:

For any prime number p at least one element of the set $\{0, 1, \ldots, p - 1\}$ does not occur more than once among the remainders $r_1(p), r_2(p), \ldots, r_n(p)$.

We conclude this chapter with a statement on large gaps in the set of visible points which can be proved by using the Chinese remainder theorem.

Recall that a lattice point in the two-dimensional cartesian xOy coordinate system is a point with integer coordinates x, y. A lattice point $P(x, y)$ is visible from the origin $O(0, 0)$ of the system if there is no lattice point inside the straight line segment OP. The lattice point $P(x, y)$ is visible from O if and only if its coordinates are coprime.

Sets of visible lattice points were studied in Section 3.3 of Chapter III. Here we shall show the following.

The set of lattice points in the plane, visible from the origin O of the coordinate system, contains arbitrarily large square gaps.

More precisely:

Problem 64

Given any natural number n, there exists a lattice point (a, b) such that none of the lattice points $(a + r, b + s)$, where $0 < r \leqslant n$, $0 < s \leqslant n$, is visible from the origin (Fig. 1.31). Prove this assertion.

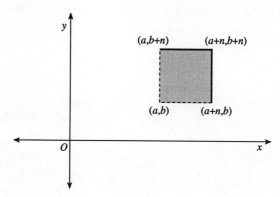

Fig. 1.31

Hint. Consider the sequence of all prime numbers:

$$p_1 = 2, \quad p_2 = 3, \quad p_3 = 5, \quad p_4, \quad p_5, \ldots$$

Use the first n^2 prime numbers to construct the $n \times n$ matrix

$$M = \begin{pmatrix} p_1 & p_2 & \cdots & p_n \\ p_{n+1} & p_{n+2} & \cdots & p_{2n} \\ p_{2n+1} & p_{2n+2} & \cdots & p_{3n} \\ p_{n^2-n+1} & p_{n^2-n+2} & \cdots & p_{n^2} \end{pmatrix}$$

Form the product m_i of the entries in the ith (horizontal) row of M and the product M_j of the entries in the jth (vertical) column of M for $i, j = 1, 2, \ldots, n$, and consider the two systems of linear congruences:

(I) $\begin{cases} x \equiv -1 \,(\mathrm{mod}\ m_1) \\ x \equiv -2 \,(\mathrm{mod}\ m_2) \\ \cdots \ \cdots \ \cdots \ \cdots \\ x \equiv -n \,(\mathrm{mod}\ m_n) \end{cases}$ and (II) $\begin{cases} y \equiv -1 \,(\mathrm{mod}\ M_1) \\ y \equiv -2 \,(\mathrm{mod}\ M_2) \\ \cdots \ \cdots \ \cdots \ \cdots \\ y \equiv -n \,(\mathrm{mod}\ M_n) \end{cases}$

Show that the systems (I) and (II) have unique respective solutions a and b modulo $m_1 m_2 \cdot \ldots \cdot m_n = M_1 M_2 \cdot \ldots \cdot M_n$, and verify that the square with vertices (a, b), $(a+n, b)$, $(a+n, b+n)$, and $(a, b+n)$ contains no visible lattice points, neither in its interior nor on its 'upper' and 'right-hand side' edges.

VI Two famous inequalities and some related problems

Introduction

We shall focus attention on two classical inequalities, with various applications in modern mathematics:

(a) *Cauchy's inequality*, which compares the geometric mean $\sqrt[n]{a_1 a_2 \cdot \ldots \cdot a_n}$ with the arithmetic mean $(a_1 + a_2 + \ldots + a_n)/n$ of arbitrary, non-negative real numbers $a_1, a_2, \ldots, a_n$:

$$\sqrt[n]{a_1 a_2 \cdot \ldots \cdot a_n} \leqslant \frac{a_1 + a_2 + \ldots + a_n}{n}; \qquad (1.22)$$

and

(b) the *isoperimetric theorem* for plane figures which states that
of all plane figures with a given perimeter, the circle has the greatest area.

In Section 6.1 we deduce (1.22) from another inequality (1.24), which is interesting for its own sake. In Section 6.2 we generalize Cauchy's inequality. This generalization is shown to be a source of a wide range of seemingly unrelated inequalities.

The isoperimetric theorem has a long history. It was formulated more than 2000 years ago; however its first rigorous proof was carried out as late as the nineteenth century. All known proofs of the isoperimetric theorem are far from elementary and cannot be presented in this book. We shall limit ourselves to giving a brief outline of the ideas behind the proof suggested by Jacob Steiner (1796–1863) and corrected by Karl Weierstrass (1815–1897). This will be done in Section 6.3. This section contains a number of problems dealing with special cases of the isoperimetric theorem.

6.1   A proof of Cauchy's theorem

We shall start with the following problem.

Problem 65

Let $x_1, x_2, \ldots, x_n$ and $y_1, y_2, \ldots, y_n$ be real numbers such that $x_1 \leqslant x_2 \leqslant \ldots \leqslant x_n$ and $y_1 \leqslant y_2 \leqslant \ldots \leqslant y_n$.

51

For an arbitrary permutation $y_{i_1}, y_{i_2}, \ldots, y_{i_n}$ of $y_1, y_2, \ldots, y_n$ consider the sum $x_1 y_{i_1} + x_2 y_{i_2} + \ldots + x_n y_{i_n}$. (Recall that a permutation of $y_1, y_2, \ldots, y_n$ is an arrangement of the terms of this sequence in a particular order.)
Prove that

$$x_1 y_{i_1} + x_2 y_{i_2} + \ldots + x_n y_{i_n} \leqslant x_1 y_1 + x_2 y_2 + \ldots + x_n y_n \qquad (1.23)$$

and

$$x_1 y_{i_1} + x_2 y_{i_2} + \ldots + x_n y_{i_n} \geqslant x_1 y_n + x_2 y_{n-1} + \ldots + x_n y_1 \qquad (1.24)$$

The above inequalities have many useful applications. One of them is a proof of inequality (1.22), named after Augustin Louis Cauchy (1789–1851). (There are various proofs of (1.22), see for example Dörrie [19].) This is the subject of Problem 66.

Problem 66

Let $a_1, a_2, \ldots, a_n$ be n arbitrary, positive real numbers where $n \in \{1, 2, 3, \ldots\}$.
Construct the numbers

$$x_i' = \frac{a_1 a_2 \cdot \ldots \cdot a_i}{(\sqrt[n]{a_1 a_2 \cdot \ldots \cdot a_n})^i} \quad \text{for } i = 1, \ldots, n,$$

and their reciprocals $y_i' = 1/x_i'$.

(a) Applying (1.24) to x_i' and y_i' prove that

$$\underbrace{1 + 1 + \ldots + 1}_{n} \leqslant \frac{a_1}{\sqrt[n]{a_1 a_2 \cdot \ldots \cdot a_n}} + \frac{a_2}{\sqrt[n]{a_1 a_2 \cdot \ldots \cdot a_n}} + \ldots +$$

$$+ \frac{a_n}{\sqrt[n]{a_1 a_2 \cdot \ldots \cdot a_n}},$$

leading to (1.22).

(b) When does equality hold in (1.22)?

Cauchy's inequality can be interpreted geometrically: e.g. for $n = 2$ and for $n = 3$ the numbers in the inequality can be taken as the length of line segments in the plane, respectively in three-dimensional space. This leads to:

Problem 67

Prove special cases of (1.22) for $n = 2$ and $n = 3$ by using geometrical arguments.

6.2   A generalization of Cauchy's inequality and its applications

The following intriguing generalization of Cauchy's inequality is given in Shleifer [21].

Problem 68

Let a_{ij}, where $i = 1, 2, \ldots, n$ and $j = 1, 2, \ldots, k$ for some natural numbers n and k, be non-negative real numbers, arranged in an $n \times k$ matrix M shown below:

$$
M = \begin{pmatrix} a_{11} & a_{12} & \ldots\ldots & a_{1k} \\ a_{21} & a_{22} & \ldots\ldots & a_{2k} \\ \ldots & \ldots & \ldots\ldots & \\ a_{n1} & a_{n2} & \ldots\ldots & a_{nk} \end{pmatrix} \begin{matrix} \to G_1 \\ \to G_2 \\ \\ \to G_n \end{matrix}
$$

$$
\begin{matrix} \downarrow & \downarrow & & \downarrow \\ A_1 & A_2 & \ldots\ldots & A_k \end{matrix}
$$

Denote by G_i the geometric mean of the entries in the ith row and by A_j the arithmetic mean of the entries in the jth column of M, that is

$$
G_i = \sqrt[k]{a_{i1}\, a_{i2} \cdot \ldots \cdot a_{ik}} \qquad \text{for } i = 1, 2, \ldots, n
$$

and

$$
A_j = \frac{a_{1j} + a_{2j} + \ldots + a_{nj}}{n} \qquad \text{for } j = 1, 2, \ldots, k
$$

(a) Prove that

$$
\sqrt[k]{A_1 A_2 \cdot \ldots \cdot A_k;} \geqslant \frac{G_1 + G_2 + \ldots + G_n}{n} \tag{1.25}
$$

(*Hint*. Distinguish two cases: case 1, when $A_i = 0$ for at least one $i \in \{1, 2, \ldots, k\}$; case 2, when $A_i > 0$ for all $i \in \{1, 2, \ldots, k\}$. In the latter case apply (1.22) to $a_{i1}/A_1, a_{i2}/A_2, \ldots, a_{ik}/A_k$.)

(b) Under which conditions does equality hold in (1.25)?

By assigning special values to n, k and to the entries a_{ij} of the matrix M we obtain various special cases of (1.25). Some of them represent well known relations. A few examples are given below. First of all we shall show that (1.25) is indeed a generalization of (1.22).

Problem 69

For n non-negative real numbers $a_1, a_2, \ldots, a_n$ construct the matrix

$$
M = \begin{pmatrix}
a_1 & a_2 & \ldots\ldots & a_{n-1} & a_n \\
a_2 & a_3 & \ldots\ldots & a_n & a_1 \\
a_3 & a_4 & \ldots\ldots & a_1 & a_2 \\
\ldots & \ldots & \ldots\ldots & & \\
a_n & a_1 & \ldots\ldots & a_{n-2} & a_{n-1}
\end{pmatrix}
$$

Show that (1.25) reduces to (1.22) in this special case.

Problem 70

As a special case of (1.25) prove the 'Cauchy–Bunyakowski inequality'

$$
(a_1^2 + a_2^2 + \ldots + a_n^2)(b_1^2 + b_2^2 + \ldots + b_n^2) \geqslant (a_1 b_1 + a_2 b_2 + \ldots + a_n b_n)^2,
$$

where $a_1, a_2, \ldots, a_n$ and $b_1, b_2, \ldots, b_n$ are arbitrary real numbers.

Problem 71

Deduce from (1.25) the following inequality of Hölder.

Let $a_{11}, a_{21}, \ldots, a_{n1}; a_{12}, a_{22}, \ldots, a_{n2}; \ldots; a_{1k}, a_{2k}, \ldots, a_{nk}$ be k sequences of non-negative real numbers, and let $\alpha_1, \alpha_2, \ldots, \alpha_k$ be positive rational numbers such that $\alpha_1 + \alpha_2 + \ldots + \alpha_k = 1$.

Prove that

$$
a_{11}^{\alpha_1} a_{12}^{\alpha_2} \cdot \ldots \cdot a_{1k}^{\alpha_k} + a_{21}^{\alpha_1} a_{22}^{\alpha_2} \cdot \ldots \cdot a_{2k}^{\alpha_k} + \ldots + a_{n1}^{\alpha_1} a_{n2}^{\alpha_2} \cdot \ldots \cdot a_{nk}^{\alpha_k}
$$

$$
\leqslant (a_{11} + a_{21} + \ldots + a_{n1})^{\alpha_1} (a_{12} + a_{22} + \ldots + a_{n2})^{\alpha_2} \ldots
$$

$$
(a_{1k} + a_{2k} + \ldots + a_{nk})^{\alpha_k}.
$$

(*Hint.* Denote the common denominator of $\alpha_1, \alpha_2, \ldots, \alpha_k$ by D and rewrite the fractions α_i in the form $\alpha_i = E_i/D$. The numbers $E_1, E_2, \ldots, E_k$ are non-negative integers with sum $E_1 + E_2 + \ldots + E_k = D$. Construct a matrix M of n rows and D columns as shown below:

$$
M = \begin{pmatrix}
\underbrace{a_{11}\, a_{11} \ldots a_{11}}_{E_1} & \underbrace{a_{12}\, a_{12} \ldots a_{12}}_{E_2} & \ldots\ldots & \underbrace{a_{1k}\, a_{1k} \ldots a_{1k}}_{E_k} \\
a_{21}\, a_{21} \ldots a_{21} & a_{22}\, a_{22} \ldots a_{22} & \ldots\ldots & a_{2k}\, a_{2k} \ldots a_{2k} \\
\ldots\ldots\ldots\ldots\ldots & \ldots\ldots\ldots\ldots & \ldots\ldots & \ldots\ldots\ldots\ldots \\
a_{n1}\, a_{n1} \ldots a_{n1} & a_{n2}\, a_{n2} \ldots a_{n2} & \ldots\ldots & a_{nk}\, a_{nk} \ldots a_{nk}
\end{pmatrix}
$$

Apply (1.25) to this M.

Problem 72

Prove the inequality

$$\left(\frac{a_1^k + a_2^k + \ldots + a_n^k}{n}\right)^{1/k} \geq \left(\frac{a_1^m + a_2^m + \ldots + a_n^m}{n}\right)^{1/m}$$

where $a_1, \ldots, a_n$ are non-negative real numbers and k and m are integers such that $k \geq m > 0$.

(*Hint.* Form a matrix M of n rows and k columns such that the first m columns are all of the form

$$a_1^k$$
$$a_2^k$$
$$\vdots$$
$$a_n^k$$

while the entries in the remaining columns are all equal to 1).

The next problem involves an inequality related to the number e.

Problem 73

Let n be a natural number and let x be a real number such that $x \geq -n$.

(a) Prove that

$$\left(1 + \frac{x}{n+1}\right)^{n+1} \geq \left(1 + \frac{x}{n}\right)^n \qquad (1.26)$$

(*Hint.* Construct an $(n+1) \times (n+1)$ matrix $M = (a_{ij})$ with $a_{11} = a_{22} = \ldots = a_{n+1\,n+1} = 1$ and $a_{ij} = 1 + x/n$ for $i \neq j$, $i, j = 1, 2, \ldots, n+1$. Deduce (1.26) from (1.25).)

(b) Applying (1.26) prove that

$$\left(1 + \frac{1}{n}\right)^n < \left(1 + \frac{1}{n+1}\right)^{n+1} \qquad \text{for} \quad n = 1, 2, \ldots$$

and that

$$\left(1 + \frac{1}{n}\right)^{n+1} > \left(1 + \frac{1}{n+1}\right)^{n+2} \qquad \text{for} \quad n = 1, 2, \ldots$$

Remark. Both sequences $(1 + 1/1), (1 + 1/2)^2, \ldots, (1 + 1/n)^n, \ldots$ and $(1 + 1/1)^2, (1 + 1/2)^3, \ldots, (1 + 1/n)^{n+1}$ converge to e.

6.3   Special cases of the isoperimetric theorem

Two plane figures are called *isoperimetric* if they have the same perimeter. Since antiquity, mathematicians have been interested in comparing areas of isoperimetric shapes. While the isoperimetric theorem in its general form is a statement about all possible plane shapes with a given perimeter, its special cases refer only to certain classes of shapes. For example, Euclid in his famous book the *Elements* in ca. 600 BC considered the isoperimetric theorem for rectangles.

Problem 74

Prove that of all rectangles with a given perimeter the square has the largest area.

There are analogous statements for isoperimetric triangles, or quadrilaterals.

Problem 75

Prove that of all triangles with a given perimeter the equilateral triangle has the largest area.

Problem 76

Prove that of all quadrilaterals with a given perimeter the square has the largest area.

The next problem compares squares with equilateral triangles:

Problem 77

Prove, by using geometrical arguments, that a square has a larger area than an equilateral triangle of the same perimeter.

The eminent nineteenth-century geometer Jacob Steiner studied arbitrary plane shapes with a given perimeter p. He proved the following.

(A) If S is any shape of perimeter p which is not a circle, then its area is less than the area of the circle with the same perimeter.

Steiner believed that (A) implies the isoperimetric theorem:

(B) Of all plane shapes with a given perimeter the circle has the largest
area.

Karl Weierstrass, a famous contemporary of Steiner, pointed out a
gap in Steiner's reasoning: (B) cannot be deduced from (A) alone; in
addition to (A) one has to prove that

(C) Among all shapes of given perimeter p *there exists* a shape S^* whose
area is larger than the area of any other shape with the same
perimeter.

Weierstrass was able to prove (C).
   (A) and (C) imply the validity of (B).

Problem 78

Assuming the validity of the isoperimetric theorem, prove that among all
n-gons with equal sides of length a the regular n-gon has the largest area.

Problem 79

Let n be a natural number and let α be an angle less than $180°/(n-1)$.

(a) Prove that

$n-1$ rhombuses of sidelength 1, with acute angles α

$n-2$ "   "   "   1, "   "   " 2α

   .   .   .   .

$n-i$ "   "   "   1, "   "   " $i\alpha$

   .   .   .   .

1   rhombus "   "   1, "   "   " $(n-1)\alpha$

can be fitted together, without gaps and overlapping, into a $2n$-gon with
parallel opposite sides (see Problem 30).

(b) Using (a) and Problem 78 find the largest value of the sum

$$s = (n-1)\sin\alpha + (n-2)\sin 2\alpha + \ldots + \sin (n-1)\alpha.$$

VII 'Mysteries' of the third dimension: on cubes, pyramids, and spheres

Introduction

(I)   Is it possible to double the cube — that is, to construct with compasses and straightedge the edge of a cube the volume of which is twice that of a given cube?

(II) Are any two pyramids with congruent bases and congruent corresponding heights equidecomposable? (In other words, is it possible to dissect one of them into finitely many polyhedra, which can be assembled to form the other pyramid?)

(III) Can thirteen congruent solid spheres touch simultaneously a given sphere of the same size?

The above questions became famous because they occupied mathematicians for centuries; finally they could be answered — each of them with a 'no'.

Three-dimensional space still abounds in 'mysterious questions', awaiting answers.

The aim of this chapter is to describe briefly the history of questions (I)–(III) and to provide a selection of problems on three-dimensional solids.

7.1   Cubes, rhombic dodecahedra, and honeybee cells

According to an ancient Greek legend, a plague devastated the island of Delos. The people of the island appealed to the oracle in Delphi for advice. They were told that the epidemic would end if the cubical altar of Apollo were doubled without change of shape.

Apollo's command was to 'double the cube' — that is to construct a cube with volume double that of the original. The ancient Greeks expressed this request in the form of the following mathematical problem, nowadays known as the *Delian problem*:

Using a straightedge and a pair of compasses construct the edge of a cube C'
whose volume is twice the volume of a cube C with edge length 1.

The edge length of the required cube C' is $\sqrt[3]{2}$. Indeed,

$$\text{Volume } C' = 2 \cdot \text{Volume } C = 2 \cdot 1^3.$$

The mathematicians of antiquity did not manage to construct a straight line segment of length $\sqrt[3]{2}$, nor did their successors. Finally, in the first half of the nineteenth century, it was proved, thanks to modern algebra, that the construction *cannot* be carried out with the prescribed tools. Modern algebra made it possible to show that the only line segments with irrational length x which can be constructed with straightedge and compasses are those for which x is a square root, a combination of square roots, or an iteration of square roots.

Attempts to double the cube led to various discoveries; one of them is the curve invented by Diocles around 180 BC, enabling him to represent graphically $\sqrt[3]{2}$. The curve, called the *cissoid of Diocles* is introduced in the next problem.

Problem 80

Let c_1 be a circle in a fixed position, OA a fixed diameter of c_1, and c_2 the tangent of c_1 at A. Consider a variable straight line ℓ through O, intersecting c_1 and c_2 at P_1 and P_2 respectively (Fig. 1.32).

The locus of P on ℓ such that $OP = OP_2 - OP_1$ is called the cissoid of Diocles.

(a) Set up a cartesian coordinate system with O as origin and with A on the positive x-axis (Fig. 1.32). Show that the equation of the curve in this coordinate system is

$$y^2 = \frac{x^3}{2a - x},$$

where a is the radius of c_1.

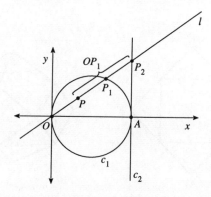

Fig. 1.32

(b) Mark the point B $(O, 4a)$ on the y-axis and join it to A. Denote the intersection of BA with the cissoid by P^*, and the intersection of OP^* with c_2 by P_2^*. Prove that $(AP_2^*)^3 = 2(OA)^3$. Thus, if $OA = 1$, the length of AP_2^* is $\sqrt[3]{2}$.

Thus, Diocles did double the cube — although not in the classical sense.

(c) In the seventeenth century Sir Isaac Newton (1642–1727) showed that the cissoid of Diocles can be generated with a carpenter's square (Fig. 1.33) as follows.

Let the outside edge of the square be ACB, *with AC as the shorter arm*. Draw a straight line ℓ and mark a point R at distance AC from ℓ. Move the square so that A always lies on ℓ and BC always passes through R. Show that the midpoint P of AC describes a cissoid of Diocles.

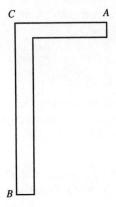

Fig. 1.33

The doubling of a square can be performed by the following simple construction (Fig. 1.34(a)).

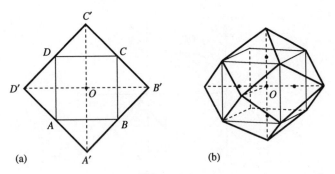

Fig. 1.34

Reflect the centre O of a square $ABCD$ in each of its sides AB, BC, CD, and DA to obtain the points A', B', C', and D' respectively. Join A' to A and B, B' to B and C, C' to C and D, and D' to D and A. The resulting polygon is a square $A'B'C'D'$ whose area is double that of $ABCD$.

A similar construction, performed on a cube C leads to a solid whose volume is double that of C:

Reflect the centre of C in each face of C and join each newly obtained point to the four nearest vertices of C. It is not difficult to show that the 24 line segments obtained in this way form 12 congruent rhombuses (Fig. 1.34(b)).

The solid, bounded by the 12 congruent rhombuses is called a *rhombic dodecahedron*. It was discovered by Johannes Kepler (1571–1630) who proved that congruent copies of rhombic dodecahedra tessellate space (that is, congruent rhombic dodecahedra can be placed next to one another without gaps and overlapping to fill space) and noticed a similarity between these shapes and the honeybee cells. Kepler knew that a honeybee cell is not a hexagonal prism, as some of us would have thought. It has a regular hexagonal base with six edges perpendicular to it; however, these edges alternate in length so that the lateral faces of the cell are congruent trapezoids. This construction is closed with a 'roof' consisting of three congruent rhombuses (Fig. 1.35(a)).

The cells of a honeycomb form two layers, one of them standing 'upright' and the other 'upside down' so that the roofs of one layer fit tightly between the roofs of the other (Fig. 1.35(b)). This property of the

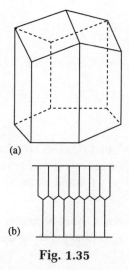

(a)

(b)

Fig. 1.35

roofs — to tessellate space — made Kepler think that the rhombuses of the roof had the same angles as the rhombuses of the rhombic dodecahedra.

Scientists have later verified that Kepler's conjecture was correct. René Réaumur (1683–1757) wondered why all honeybee cells were covered with the same type of rhombus, and came up with the idea that the bees' intention is to save on building material. He asked König, a contemporary Swiss mathematician, to solve the following problem.

Problem 81

(a) Close a regular hexagonal prism with a roof consisting of three congruent rhombuses so that the newly obtained solid S has a prescribed volume and a minimal surface area.

(b) Show that the angles of the rhombuses of S are equal to the angles of the rhombuses in a rhombic dodecahedron.

Now a few problems about cubes:

Problem 82

Show that it is possible to cut a hole in a cube, through which another cube of equal size can pass.

Problem 83

Dissect a cube into three congruent pyramids.

Problem 84

A cube $ABCDA'B'C'D'$ is rotated through 60° about its diagonal BD'. If its edges are of length a, calculate in terms of a the volume of the solid which is the intersection of the original cube and of its rotated image.

Many problems of two-dimensional geometry can be solved in a simple and elegant way by applying arguments from three-dimensional geometry. Here is an example.

Problem 85

Figure 1.36 shows a regular hexagon *ABCDEF* in the plane with points P_1, P_2, P_3, P_4, P_5 and P_6 on its sides *AB, BC, CD, DE, EF,* and *FA* respectively.

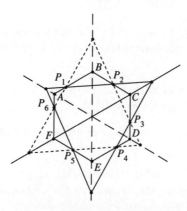

Fig. 1.36

If the straight lines $P_1 P_2$ and $P_3 P_4$ meet on the straight line *FC*,

    "   "   " $P_3 P_4$ " $P_5 P_6$ " " "   "   " *BE*,

and "   "   " $P_5 P_6$ " $P_1 P_2$ " " "   "   " *DA*,

then

    the straight lines $P_2 P_3$ and $P_4 P_5$ meet on the straight line *AD*,

    "   "   " $P_4 P_5$ " $P_6 P_1$ " " "   "   " *CF*

and "   "   " $P_6 P_1$ " $P_2 P_3$ " " "   "   " *EB*.

Prove this statement — using properties of a cube!

7.2   Tetrahedra — compared with triangles

Three non-collinear points can be considered as the vertices of a triangle, while four non-coplanar points can be viewed as the vertices of a tetrahedron.

Thus, tetrahedra are three-dimensional generalizations of triangles. This raises the question: which properties of tetrahedra are analogous, and which are different from the corresponding properties of triangles?

Some answers to the above question are given in this section. We start by recalling a few well known properties of triangles (for proofs see for example Coxeter and Greitzer [26]).

In any triangle:

(1) the perpendicular bisectors of the sides meet in a common point; this point is the centre of a circle passing through all vertices of the triangles;

(2) the medians, that is the straight line segments joining the vertices to the midpoints of the opposite sides, meet in a common point — the *centroid* of the triangle;

(3) the centroid divides each median in the ratio 2:1 (starting from the vertex of the triangle on a median);

(4) the altitudes meet in a common point, called the *orthocentre* of the triangle;

(5) the midpoints of the sides, the feet of the altitudes, and the midpoints of the segments joining the vertices to the orthocentre lie on a common circle, called the *nine-point circle* of the triangle;

(6) there are four circles, touching the three straight lines, which carry the sides of the triangle; one of them is inside, the three other circles are outside the triangle.

Properties (1), (2), and (3) have three-dimensional analogues:

Problem 86

Prove that in *any* tetrahedron the following statements are true.

(a) The six planes, each of them passing through the midpoint of one of the six edges, and perpendicular to this edge, meet in a common point. This point is the centre of the sphere passing through all vertices of the tetrahedron.

(b) The four straight line segments joining the vertices with the centroids of the opposite faces pass through a common point — the *centroid* of the tetrahedron.

(c) The centroid divides each straight line segment, joining a vertex to the centroid of the opposite face, in the ratio 3:1.

On the other hand, not every tetrahedron has an *orthocentre*, that is a point common to all of its altitudes. For example, in the tetrahedron $ABCG$, forming part of the cube $ABCDEFGH$ in Fig. 1.37, the altitudes GC and AB are skew.

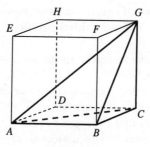

Fig. 1.37

The next problem gives conditions which are necessary and sufficient for the existence of the orthocentre in a tetrahedron.

Problem 87

Prove that a tetrahedron T has an orthocentre if and only if anyone of the following conditions holds:

(a) its opposite edges are perpendicular;

(b) one of its altitudes passes through the orthocentre of the corresponding base;

(c) the straight line segments connecting the midpoints of opposite edges are of equal length.

The nine-point circle of a triangle has the following analogue.

Problem 88

Prove that if a tetrahedron has an orthocentre then the nine-point circles of its six faces belong to a common sphere (the '24 point sphere').

The last of the properties (1)–(6) which we have listed for triangles has an intriguing counterpart for tetrahedra.

Problem 89

Prove that there are at least five and at most eight spheres touching the four planes which carry the faces of *any* tetrahedron.

Now we state a property common to all tetrahedra.

Problem 90

Prove that the six planes, each of which passes through the midpoint of an edge of an arbitrary tetrahedron and is perpendicular to the opposite edge, meet in a common point (known as the point of Monge).

7.3   Are shapes of equal 'content' equidecomposable?
Hilbert's third problem

The 'content' of a plane polygon is its area, and the 'content' of a polyhedron is its volume.

Two polygons are called *equidecomposable* if one of them can be dissected into *finitely* many polygons which fit together (without gaps and overlapping) to form the other polygon. Similarly, two polyhedra and *equidecomposable* if one of them can be dissected into *finitely* many polyhedra which fit together (without gaps and overlapping) to form the other polyhedron.

Equidecomposable shapes have equal content. A simple application of this fact led the ancient Greeks to the formula about the area of a triangle:

An arbitrary triangle with base b and corresponding height h and a rectangle with base b and height h/2 are equidecomposable; hence the area A of the triangle is equal to bh/2.

An analogous formula for the volume of a pyramid was also known in antiquity. The volume of a pyramid is equal to $BH/3$ where B is the area of the base, and H is the corresponding height of the pyramid. However, all proofs of this formula, starting with that given by Archimedes of Syracuse (287–212 BC), were based on infinitesimal processes instead of on arguments based on equidecomposability.

By the end of the nineteenth century many mathematicians suspected that pyramids of base B and height H are not necessarily equidecomposable with prisms of base congruent to B, and of height $H/3$. David Hilbert (1862–1943) was convinced that a thorough study of equidecomposable solids of equal content would have an impact on further developments in mathematics. In his famous lecture at the Second International Mathematical Congress, held in 1900 in Paris, Hilbert posed 23 problems which he considered to be of fundamental importance in mathematics. The third problem was the following:

Find two pyramids with the same face as base and of equal height which are not equidecomposable.

Only a few months later, Max Dehn gave an elegant solution of Hilbert's third problem. Dehn's solution consists of two steps:

First he established a *necessary* condition ($\mathbb{C}$) which must be satisfied by any pair of equidecomposable polyhedra; then he gave an example of two pyramids with a common base and equal heights which did not satisfy condition ($\mathbb{C}$). Hence the two pyramids were not equidecomposable.

In this section, following Dehn's ideas, we shall give a solution to Hilbert's problem on pyramids.

Dehn's condition ($\mathbb{C}$) is about dihedral angles of polyhedra. A *dihedral angle* of a polyhedron is the angle between two of its faces through a common edge. Dihedral angles will be measured here in radians. Dehn has proved the following:

Theorem D. Let P and Q be two equidecomposable polyhedra with dihedral angles $\alpha_1, \alpha_2, \ldots, \alpha_n$ and $\beta_1, \beta_2, \ldots, \beta_m$ respectively. Then there exist non-negative integers $a_1, a_2, \ldots, a_n$ and $b_1, b_2, \ldots, b_m$ and an integer k such that

$$(a_1 \alpha_1 + a_2 \alpha_2 + \ldots + a_n \alpha_n) - (b_1 \beta_1 + b_2 \beta_2 + \ldots + b_m \beta_m) = k\pi \quad (\mathbb{C})$$

Proof of Theorem D. Since P and Q are equidecomposable, both can be constructed from the same set of finitely many polyhedra $\Pi_1, \Pi_2, \ldots, \Pi_r$ as building blocks by putting them together in two (mostly different) ways.

Dehn's idea was to calculate the sum Σ of the dihedral angles of all building blocks $\Pi_1, \Pi_2, \ldots, \Pi_r$ in two different ways.

Way 1. Consider $\Pi_1, \Pi_2, \ldots, \Pi_r$ fitted together to form P. On the surface of each Π_i mark — say in red — those edges or segments of edges of the neighbouring blocks which lie on Π_i. If Π_i has edges which are edges, or parts of edges, of P, mark them also in red. In this way a network of red-line segments is produced. Each red-line segment belongs to the edge of one or more dihedral angles of one or more building blocks $\Pi_1, \Pi_2, \ldots, \Pi_r$.

In order to determine Σ we shall distinguish three types of red-line segments in the network.

Type I: red-line segments on an edge of a dihedral angle α_j or P.

The dihedral angles of the building blocks, which belong to the edge of α_j, add up to a multiple $a_j \alpha_j$ of α_j. Since j varies from 1 to n, the sum of all dihedral angles of $\Pi_1, \Pi_2, \ldots, \Pi_r$ with edges of type I is

$$\Sigma' = a_1 \alpha_1 + a_2 \alpha_2 + \ldots + a_n \alpha_n.$$

Type II: red-line segments inside a face of a building block of P.

The dihedral angles of the building blocks with such an edge add up to π; thus if the network contains s segments of type II, then all dihedral angles with their edges in the building blocks $\Pi_1, \Pi_2, \ldots, \Pi_r$ add up to $\Sigma'' = s\pi$.

Type III: all remaining line segments of the network.

They are inside P forming an edge, or part of an edge, of each of the building blocks to which they belong. The dihedral angles of the blocks with such an edge add up to 2π; hence if there are t segments of type III in the network, the total of the corresponding dihedral angles of $\Pi_1, \Pi_2, \ldots, \Pi_r$ is $\Sigma''' = t \cdot 2\pi$.

In view of the above arguments Σ can be expressed in the form $a_1\alpha_1 + a_2\alpha_2 + \ldots + a_n\alpha_n + s\pi + 2t\pi$, that is

$$\Sigma = a_1\alpha_1 + a_2\alpha_2 + \ldots + a_n\alpha_n + N_1\pi \qquad (1.27)$$

where $a_1, a_2, \ldots, a_n$ and N_1 are non-negative integers.

Way 2. This consists in fitting together the same blocks $\Pi_1, \Pi_2, \ldots, \Pi_r$ to form Q. In this arrangement the edges, or segments of edges, of building blocks lying on the surface of a block Π_i, marked — say green — lead to a 'green' network which is, in general, different from the red network considered previously. The same counting arguments as in the case of red-line segments lead to an expression for Σ analogous to (1.27):

$$\Sigma = b_1\beta_1 + b_2\beta_2 + \ldots + b_m\beta_m + N_2\pi \qquad (1.28)$$

where $b_1, b_2, \ldots, b_m$ and N_2 are non-negative integers.

Relations (1.27) and (1.28) together yield Dehn's condition $(\mathbb{C})$. This finishes the proof of Theorem D.

We are now able to show that there are pyramids with the same base and equal height which are not equidecomposable:

Problem 91

Let P be a pyramid with square base of side length a and with height of length $a/2$ erected above the centre of the base (Fig. 1.38(a)). If Q is a pyramid with a square base of side length a and height of length $a/2$ erected above the midpoint of an edge of the base (Fig. 1.38(b)) prove that P and Q do not satisfy condition $(\mathbb{C})$.

By solving Problem 91 we solve Hilbert's Third Problem.

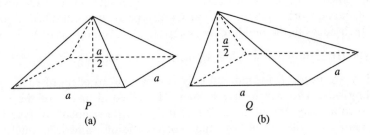

Fig. 1.38

Problem 92

Find out whether a regular tetrahedron and a cube of the same volume are equidecomposable.

Problem 93

Is it possible to dissect the pyramid P described in Problem 91 into finitely many polyhedra which can be assembled to form a cuboid having the same base as P?

7.4 Problems on spheres

In 1694 Isaac Newton (1642–1727) and David Gregory (1661–1708) discussed the distribution of stars of various magnitudes and disagreed over the following question:

What is the greatest number N of non-overlapping, congruent spheres, which can touch simultaneously a sphere of the same size?

Gregory though that $N = 13$, Newton believed that $N = 12$. It took 180 years to settle the dispute, when R. Hoppe proved that Newton was right. For a proof of Hoppe's result readers are referred to Wassermann [30]. Here we shall consider the problem of 12 spheres touching a 13th sphere.

Problem 94

Construct 12 non-overlapping congruent spheres, all of them touching a 13th sphere of the same size.

'Sphere-packing' problems — on arrangements of non-overlapping, congruent spheres, touching one another — constitute a flourishing field of modern mathematics; it abounds in solved and unsolved problems. One of the main research tasks is to find out whether there is a 'most economical' packing, i.e. one of greatest density. (The density of a packing is $\lim_{R \to \infty} \left(n_R(4r^3\pi/3)/(4R^3\pi/3) \right)$ where r is the radius of the spheres in the packing and n_R is the number of those spheres in the packing which lie entirely within a sphere of radius R.)

We end this chapter by listing a few problems about arcs of great circles of a sphere.

Any plane meeting a sphere S in more than one point intersects it along a circle. Planes passing through the centre of S intersect it along circles having the greatest possible radii — equal to the radius of S; these circles are called the *great circles* of S. Their arcs play the role of straight line segments on the sphere:

Problem 95

Prove that the shortest curve on the surface of a sphere S joining two points P and Q on the sphere is the smaller arc of the great circle through P and Q.

Problem 96

Prove that it is not possible to draw three arcs of great circles on S, each with a central angle of $300°$, such that no two of them have a point in common.

Problem 97   *(Kvant M1020)*

Let S be a sphere of unit radius, and let p be a 'polygonal line' on S, that is a curve consisting of arcs of great circles. Prove that if the length of p is less than π (in radians), then there exists a great circle on S which does not intersect p.

VIII A glimpse of higher-dimensional spaces: hypercubes, lattice paths, and related number patterns

Introduction

The classical geometries of dimensions two and three, the elementary properties of which are taught at school, were first described systematically by Euclid, about 300 BC, in his famous book the *Elements*. For this reason these geometries are called *euclidean geometries* of dimensions two and three respectively.

Higher dimensional euclidean geometries emerged only in the second half of the nineteenth century.

Our aims in this chapter are the following:

to explain briefly what is meant by n-dimensional real euclidean space (Section 8.1),

to introduce n-dimensional hypercubes (Section 8.2), and

to study hypercubic lattice paths and related number patterns (Section 8.3).

8.1 On n-dimensional real euclidean spaces

We shall introduce n-dimensional real euclidean spaces by generalizing the concept of cartesian coordinates, used in the study of two- and three-dimensional euclidean geometries.

Figure 1.39(a) and (b) shows two rectangular cartesian coordinate systems, constructed in a euclidean real plane Σ_2 and in euclidean real space Σ_3 respectively. The two coordinate systems have the following common features.

Both have a frame of reference:

in Σ_2 the frame of reference consists of two perpendicular straight lines, the *coordinate axes* x_1 *and* x_2, meeting at a point O, the *origin* of the system.

in Σ_3 the frame of reference consists of three pairwise perpendicular planes x_1Ox_2, x_2Ox_3, and x_3Ox_1 meeting at a point O, the *origin* of the system.
x_1Ox_2 and x_2Ox_3 intersect along x_2,
x_2Ox_3 and x_3Ox_1 intersect along x_3,
x_3Ox_1 and x_1Ox_2 intersect along x_1; the straight lines x_1, x_2, x_3 are the *coordinate axes*.

In this coordinate system:
(i) each point P is represented by an ordered pair of real numbers (x_1, x_2) called the *coordinates* of P; the coordinates x_1 and x_2 of P are its distances from the x_2-axis and the x_1-axis respectively;

In this coordinate system:
(i′) each point P is represented by an ordered triple of real numbers (x_1, x_2, x_3) called the *coordinates* of P; the coordinates x_1, x_2, x_3 are the distances of P from the coordinate planes x_2Ox_3, x_3Ox_1, and x_1Ox_2 respectively;

(ii) each *straight line* is represented by an equation of the form

$$a_1 x_1 + a_2 x_2 + b = 0$$

where a_i, $i = 1, 2$, and b are fixed real numbers, and not all a_i are equal to 0;

(ii′) each *plane* is represented by an equation of the form

$$a_1 x_1 + a_2 x_2 + a_3 x_3 + b = 0$$

where a_i, $i = 1, 2, 3$, and b are fixed real numbers and not all a_i are equal to 0;

(iii) the *distance* between two points $P(x_1, x_2)$ and $P'(x_1', x_2')$ is given by the formula

$$PP' = \sqrt{(x_1 - x_1')^2 + (x_2 - x_2')^2}.$$

(iii′) the *distance* between two points $P(x_1, x_2, x_3)$ and $P'(x_1', x_2', x_3')$ is given by the formula

$$PP' = \sqrt{(x_1 - x_1')^2 + (x_2 - x_2')^2 + (x_3 - x_3')^2}$$

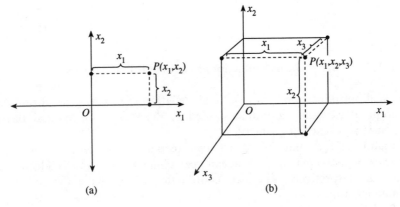

(a)   (b)

Fig. 1.39

The analogies between (i)–(iii) and (i')–(iii') suggest the following generalization of Σ_2 and Σ_3 to *n-dimensional spaces* Σ_n for any $n = 0, 1, 2, 3, 4, \ldots.$

The n-dimensional real euclidean space Σ_n, $n \geqslant 1$, consists of all ordered n-tuples of real numbers $(x_1, x_2, \ldots, x_n)$, called the points of Σ_n. The numbers x_i are the coordinates of $(x_1, x_2, \ldots, x_n)$.

The distance between two points $P(x_1, x_2, \ldots, x_n)$ and $P'(x'_1, x'_2, \ldots, x'_n)$ is defined by the formula

$$PP' = \sqrt{(x_1 - x'_1)^2 + (x_2 - x'_2)^2 + \ldots + (x_n - x'_n)^2}.$$

The role of lines, respectively planes in Σ_2 and Σ_3, is played by hyperplanes in Σ_n:

A hyperplane of Σ_n is the set of all points $(X_1, X_2, \ldots, X_n)$ whose coordinates x_i satisfy an equation of the form

$$a_1 x_1 + a_2 x_2 + \ldots a_n x_n + b = 0,$$

where $a_1, a_2, \ldots, a_n$ and b are fixed real numbers, and not all a_i are equal to 0.

The hyperplanes Π_i of Σ_n, with equations $x_i = 0$ for $i = 1, \ldots, n$ are called the *coordinate hyperplanes* of Σ_n. Their name is justified by the following property.

Problem 98

Let $P(x_1, x_2, \ldots, x_n)$ be a given point, and let Π_i be the coordinate hyperplane of Σ_n with equation $x_i = 0$. Define *the distance $d(P, \Pi_i)$ between P and Π_i* as the shortest of the distances PP', where P' is an arbitrary point of Π_i.
Prove that $d(P, \Pi_i) = |x_i|$.

Thus, the coordinate hyperplanes are generalizations of the coordinate axes and coordinate planes in two- and three-dimensional cartesian coordinate systems respectively. They constitute a *frame of reference* for Σ_n. The common point $(0, 0, \ldots, 0)$ of the coordinate hyperplanes is the *origin* in this system of reference.

The following facts about euclidean planes and 3-spaces are well known.

(iv) Any two distinct points $P'(x_1', x_2')$ and $P''(x_1'', x_2'')$ of Σ_2 are contained in a unique line ℓ of Σ_2.

ℓ consists of those points $P(x_1, x_2)$ of Σ_2 whose coordinates are expressible by the formulae

$$x_1 = x_1'' + \lambda(x_1' - x_1'')$$
$$x_2 = x_2'' + \lambda(x_2' - x_2'')$$

where λ ranges over the set $\mathbb{R}$ of real numbers.

(iv') Any two distinct points $P'(x_1', x_2', x_3')$ and $P''(x_1'', x_2'', x_3'')$ of Σ_3 are contained in a unique line ℓ of Σ_3.

ℓ consists of those points $P(x_1, x_2, x_3)$ of Σ_3 whose coordinates are expressible by the formulae

$$x_1 = x_1'' + \lambda(x_1' - x_1'')$$
$$x_2 = x_2'' + \lambda(x_2' - x_2'')$$
$$x_3 = x_3'' + \lambda(x_3' - x_3'')$$

where λ ranges over $\mathbb{R}$.

(v') Any three points $P'(x_1', x_2', x_3')$, $P''(x_1'', x_2'', x_3'')$, and $P'''(x_1''', x_2''', x_3''')$ of Σ_3, which do not belong to a common line, are contained in a unique plane of Σ_3.

π consists of those points $P(x_1, x_2, x_3)$ of Σ_3 whose coordinates are expressible by the formulae

$$x_1 = x_1''' + \lambda_1(x_1' - x_1''') + \lambda_2(x_1'' - x_1''')$$
$$x_2 = x_2''' + \lambda_1(x_2' - x_2''') + \lambda_2(x_2'' - x_2''')$$
$$x_3 = x_3''' + \lambda_1(x_3' - x_3''') + \lambda_3(x_3'' - x_3''')$$

where λ_1 and λ_2 range independently over $\mathbb{R}$.

(iv), (iv'), and (v') yield the following generalizations in Σ_n.

(a) Let $P' = (x_1', x_2', \ldots, x_n')$ and $P'' = (x_1'', x_2'', \ldots, x_n'')$ be two arbitrary distinct points of Σ_n and let $\langle P', P'' \rangle$ be the set of points $(x_1, x_2, \ldots, x_n)$ with coordinates satisfying the equations

$$x_i = x_i'' + \lambda(x_i' - x_i'') \quad \text{for} \quad i = 1, 2, \ldots, n \text{ and } \lambda \in \mathbb{R}.$$

It can be proved that

$\langle P', P'' \rangle$ is a real euclidean line, and that
$\langle P', P'' \rangle$ *is the unique one-dimensional real euclidean space in* Σ_n, *containing both* P' *and* P''.

(b) Let $P^{(1)} = (x_1^{(1)}, x_2^{(1)}, \ldots, x_n^{(1)})$, $P^{(2)} = (x_1^{(2)}, x_2^{(2)}, \ldots, x_n^{(2)}), \ldots, P^{(j)} = (x_1^{(j)}, x_2^{(j)}, \ldots, x_n^{(j)})$ be j points of Σ_n for a given natural number $j \in \{3, \ldots, n\}$.

Denote by $\langle P^{(1)}, P^{(2)}, \ldots, P^{(j)} \rangle$ the set of points whose coordinates $(x_1, x_2, \ldots, x_n)$ satisfy the equations

$$x_i = x_i^{(j)} + \lambda_1(x_i^{(1)} - x_i^{(j)}) + \lambda_2(x_i^{(2)} - x_i^{(j)}) + \ldots + \lambda_{j-1}(x_i^{(j-1)} - x_i^{(j)})$$

for $i = 1, \ldots, n$ with $\lambda_1, \lambda_2, \ldots, \lambda_{j-1}$ ranging independently over $\mathbb{R}$. If $P^{(1)}, P^{(2)}, \ldots, P^{(j)}$ do not belong to a common $(j-2)$-dimensional euclidean real space, contained in Σ_n, then $\langle P^{(1)}, P^{(2)}, \ldots, P^{(j)} \rangle$ has the structure of Σ_{j-1}. In that case,

$\langle P^{(1)}, P^{(2)}, \ldots, P^{(j)} \rangle$ *is the unique* $(j-1)$-*dimensional real euclidean space in* Σ_n, *containing* $P^{(1)}, P^{(2)}, \ldots, P^{(j)}$.

Thus Σ_n contains real euclidean spaces of dimensions $d = 1, 2, \ldots, n-1$; they are called the *d-dimensional subspaces* of Σ_n. The points of Σ_n are considered as *subspaces of dimension* 0.

We shall accept the truth of statements (a) and (b) without outlining their proofs. (Interested readers are referred to Coxeter [31].) Using (a) and (b) we are able to formulate the following *orthogonality condition* for lines in Σ_n which pass through the origin O of Σ_n.

Problem 99

Let $P(x_1, x_2, \ldots, x_n)$ and $P'(x'_1, x'_2, \ldots, x'_n)$ be points of Σ_n, for $n \geqslant 2$, different from the origin O. Denote by α the angle between $\langle O, P \rangle$ and $\langle O, P' \rangle$. (α is uniquely defined as the angle POP' in the euclidean real plane of Σ_n, containing O, P, and P'.)
  Prove that;

(a) $\alpha = 90°$, that is $\langle O, P \rangle$ and $\langle O, P' \rangle$ are *orthogonal to one another* if and
     only if

$$x_1 x'_1 + x_2 x'_2 + \ldots + x_n x'_n = 0;$$

(b) for any α such that $0 \leqslant \alpha \leqslant 180°$,

$$\cos \alpha = \frac{x_1 x'_1 + x_2 x'_2 + \ldots + x_n x'_n}{\sqrt{x_1^2 + x_2^2 + \ldots + x_n^2} \cdot \sqrt{(x'_1)^2 + (x'_2)^2 + \ldots + (x'_n)^2}}.$$

(Thus (a) is a special case of (b).)

The orthogonality of two arbitrary lines ℓ_1 and ℓ_2 of Σ_n is defined as follows.

(∗)   In the subspace of Σ_n, determined by O and ℓ_1, consider the line $\langle O, P_1 \rangle$ parallel to ℓ_1, and in the subspace of Σ_n determined by O and ℓ_2 consider the line $\langle O, P_2 \rangle$ parallel to ℓ_2. We say that ℓ_1 is orthogonal to ℓ_2 if and only if $\langle O, P_1 \rangle$ is orthogonal to $\langle O, P_2 \rangle$.

(∗∗) *A line* ℓ *of* Σ_n *is said to be orthogonal to a hyperplane* Π *of* Σ_n *if and only if* ℓ *is orthogonal to all lines of* Π.

(∗∗∗) If a hyperplane Π' of Σ_n contains a line ℓ orthogonal to a hyperplane Π, then we say that Π' *is orthogonal to* Π.

Using the above definitions, and Problem 98, one can show that the coordinate hyperplanes of Σ_n are orthogonal to one another. In other words, the hyperplanes of Σ_n form a *rectangular frame of reference*, generalizing the rectangular frames in Σ_2 and Σ_3:

Problem 100

In Σ_n consider the hyperplanes Π_i with equations $x_i = 0$, and the lines $Ox_i = \langle O, P_i \rangle$, where P_i is the point whose ith coordinate $x_i = 1$, and the remaining coordinates are $x_1 = x_2 = \ldots = x_{i-1} = x_{i+1} = \ldots = x_n = 0$.

(a) Prove that Ox_i is orthogonal to Π_i for $i = 1, 2, \ldots, n$.

(b) Deduce that Π_i is orthogonal to Π_j for any $i, j \in \{1, 2, \ldots, n\}$ such that $i \neq j$.

The lines Ox_i are called the *coordinate axes* of Σ_n.

8.2   *n*-dimensional hypercubes

Let a be a positive real number. It is well known that in Σ_3 the set of points (x_1, x_2, x_3) such that

$$0 \leqslant x_i \leqslant a \quad \text{for} \quad i = 1, 2, 3 \tag{1.29}$$

forms a cube of edge length a (Fig. 1.40).

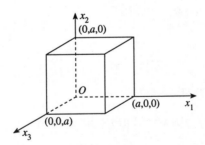

Fig. 1.40

The extension of (1.29) to n coordinates leads to *hypercubes*.

The set of points $(x_1, x_2, \ldots, x_n)$ in Σ_n such that

$$0 \leqslant x_i \leqslant a \quad \text{for} \quad i = 1, 2, \ldots, n \tag{1.30}$$

forms an *n*-dimensional hypercube of edge length a. Let us denote this hypercube by $C_n(a)$.

In general, *an n-dimensional hypercube in Σ_n is defined as a point set $C_n'(a)$ which admits a mapping μ onto $C_n(a)$ satisfying the following conditions:*

(1) *μ is one to one, and*

(2) *μ is distance preserving.*

In other words, there is a mapping μ such that every point $P' \in C_n'(a)$ has a unique image $\mu(P') \in C_n(a)$ and every point $P \in C_n(a)$ is the image of exactly one point of $C_n'(a)$. Moreover, the distance between two arbitrary points $P', Q' \in C_n'(a)$ is equal to the distance between their images $\mu(P')$ and $\mu(Q')$.

It is obvious from the definition that cubes, squares, straight-line segments, and points are three-, two-, one-, and zero-dimensional hypercubes respectively.

Our next aim is to study the number of the lower dimensional parts of the hypercubes contained in their boundaries. We start by comparing $C_3(a)$ with $C_n(a)$ for $n \geqslant 4$.

$C_3(a)$ is bounded by six planes: three planes with equations $x_i = 0$ and three planes with equations $x_i = a$.

$C_n(a)$ is bounded by $2n$ hyperplanes: n hyperplanes with equations $x_i = 0$ and n hyperplanes with equations $x_i = a$.

The intersection of a boundary plane with $C_3(a)$ is a square called a *face* of the cube.

The intersection of a boundary hyperplane with $C_n(a)$ is an $(n-1)$-dimensional hypercube called a *facet*, or a *cell* or an $(n-1)$-*dimensional part of* $C_n(a)$.

The edges of the faces of $C_3(a)$ are the *edges* of the cube.

The $(n-2)$-dimensional hypercubes, which are the facets of $C_n(a)$'s facets, are called the $(n-2)$-*dimensional parts of* $C_n(a)$.

The endpoints of the edges of $C_3(a)$ are the *vertices* of the cube.

. . . .

*The facets of the i-*dimensional parts of $C_n(a)$ are the $(i-1)$-*dimensional parts of* $C_n(a)$.

. . . .

The one-dimensional parts of $C_n(a)$ are called the *edges* of $C_n(a)$, and zero-dimensional parts of $C_n(a)$ are its *vertices*.

Problem 101

(a) For each $i \in \{0, 1, 2, 3\}$ find the number of i-dimensional parts of a four-dimensional hypercube.

(b) For each $i \in \{0, 1, \ldots, n-1\}$ find the number of i-dimensional parts of an n-dimensional hypercube.

Many questions on cubes (or other three-dimensional solids) can be answered by studying their intersections with planes, their nets or central projections. The advantage of these shapes is that their dimensions do not exceed two, and hence can be drawn in a euclidean plane. Similarly, nets and central projections of four-dimensional cubes and their intersections with three-dimensional spaces are objects in our familiar, three-dimensional euclidean space and, as such, can be visualized. This approach to $C_4(a)$ is illustrated in the last problems of this section.

Problem 102

Examine the construction of a cube's net and apply the same procedure for the construction of a net of $C_4(a)$ consisting of its facets arranged in three-dimensional space.

Problem 103

(a) Let π be a plane in Σ_3, orthogonal to the diagonal $D_3 = \langle O, (a, a, a)\rangle$ of $C_3(a)$ and passing through an arbitrary point P of D_3 (Fig. 1.41). Study the shape along which π cuts the cube $C_3(a)$ for various positions of P.

(b) Similarly, let Π be a hyperplane in Σ_4, orthogonal to the diagonal $D_4 = \langle O, (a, a, a, a)\rangle$ of $C_4(a)$ and passing through an arbitrary point P of D_4. Study the shape along which Π cuts $C_4(a)$ for various positions of P.

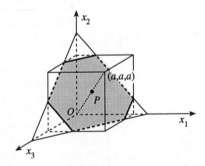

Fig. 1.41

Problem 104

In Σ_2, the *central projection* of any point Q from a point P called the *centre* onto a line $\Sigma_1 \not\ni P_1$ is the intersection Q' of PQ with Σ_1 (if such an

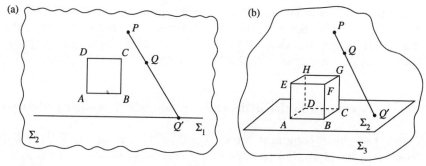

Fig. 1.42

intersection exists) (Fig. 1.42(a)). Similarly, in Σ_3 the *central projection* of any point Q from a point P — the *centre* — onto a plane $\pi_2 \not\ni P_1$ is the intersection Q' of PQ with Σ_2 (provided PQ meets Σ_2) (Fig. 1.42(b)).

(i) Draw the central projections of the square $ABCD$ onto Σ_1, and of the cube $ABCDEFGH$ onto Σ_2, depicted in Fig. 1.42.

(ii) Generalize the definition of the central projection to Σ_4 and draw a central projection of a four-dimensional hypercube onto a 3-space, using the same method as in (i).

Needless to say, $C_4(a)$ cannot be drawn in a conventional way because it is four-dimensional. Nevertheless, mathematicians found a trick for visualizing it, as the fifth term in the sequence of drawings in Fig. 1.43: when a point $C_0(a)$ is moved along a straight line from an initial to a final position (A to A') it traces out a line segment $C_1(a)$; when $C_1(a)$ is translated from AB to $A'B'$, it traces out a square $C_2(a)$; by translating $C_2(a)$ from $ABCD$ to $A'B'C'D'$ a cube $C_3(a)$ is obtained. Similarly, $C_4(a)$ is

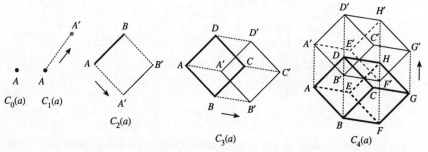

Fig. 1.43

obtained by translating $C_3(a)$ from position $ABCDEFGH$ to $A'B'C'D'E'F'G'H'$; hence $C_4(a)$ is depicted by showing the cube $ABCDEFGH$, its translate $A'B'C'D'E'F'G'H'$, and the edges of $C_4(a)$ connecting the vertices of $ABCDEFGH$ to their translates.

8.3   Hypercubic lattice paths and corresponding number patterns

Figure 1.44 shows congruent copies S_{2,p_1,p_2} of the unit square $S_2(1)$ in Σ_2, tessellating the plane. For any pair of not necessarily distinct integers p_1, p_2 the square S_{2,p_1,p_2} is defined as the set of points (x_1, x_2) such that

$$p_i \leqslant x_i \leqslant p_i + 1 \quad \text{for} \quad i = 1, 2.$$

The tessellating squares form a *square lattice* in Σ_2. The vertices of the squares are called the lattice points, and the squares are the *cells* of the lattice. Denote this lattice by L_2.

A *lattice path*, connecting two lattice points P and Q, is any polygonal line with endpoints P and Q whose sides are edges of cells. (Figure 1.44 shows one of the many lattice paths connecting a point P to a point Q.)

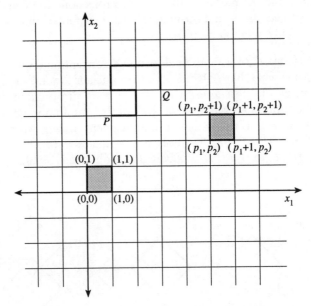

Fig. 1.44

Hypercubic lattices in Σ_n, $n \geqslant 1$, are straightforward generalizations of the square lattice in Σ_2:

For any n-tuple of not necessarily distinct integers $p_1, p_2, \ldots, p_n$ define the *hypercube* $S_{n, p_1, p_2, \ldots, p_n}$ as the set of points $(x_1, x_2, \ldots, x_n)$ in Σ_n such that

$$p_i \leqslant x_i \leqslant p_i + 1 \quad \text{for} \quad i = 1, 2, \ldots, n.$$

The set of all hypercubes $S_{n, p_1, p_2, \ldots, p_n}$ for $p_i \in \{0, \pm 1, \pm 2, \ldots\}$ tessellates Σ_n, forming a *hypercubic lattice* of Σ_n. The vertices of the hypercubes are called the *lattice points*, and the hypercubes the *cells* of the lattice. Denote this lattice by L_n.

A (hypercubic) lattice path connecting two lattice points P and Q is any polygonal line with endpoints P and Q whose sides are edges of cells.

Lattice paths play an important role in mathematical disciplines such as combinatorics, number theory, the theory of probability, and game theory. One of the reasons for this is that many problems can be formulated in terms of lattice paths (see Problem 111).

Our next aim is to point out connections between the study of lattice paths and of numbers, such as multinomial coefficients and generalized Fibonacci numbers. We shall show that:

(a) Counting of appropriate lattice paths leads to multinomial coefficients and generalized Fibonacci numbers.

(b) By counting lattice paths in different ways it is possible to obtain relations between multinomial coefficients and Fibonacci numbers. This approach often yields new proofs of well known identities, or new formulae for the numbers just mentioned above.

Before stating problems we recall that the *binomial coefficients* C_i^n, often denoted by $\binom{n}{i}$, are the numbers

$$\frac{n!}{i!(n-i)!}$$

defined for any non-negative integer n, and any integer i, with $0 \leqslant i \leqslant n$. The symbol $n!$ (called n factorial) stands for the product $n(n-1)(n-2) \cdot \ldots \cdot 3 \cdot 2 \cdot 1$ if $n \geqslant 1$; by definition, $0! = 1$.

The *trinomial coefficients*, denoted by C_{i_1, i_2, i_3}^n, are defined for any non-negative integer n and any non-negative integers i_1, i_2, i_3 with sum $i_1 + i_2 + i_3 = n$ as the numbers:

$$C_{i_1, i_2, i_3}^n = \frac{n!}{i_1! \, i_2! \, i_3!}$$

In general, the *multinomial coefficients* are defined for any non-negative integer n and arbitrary non-negative integers $i_1, i_2, \ldots, i_k$ such that $i_1 + i_2 + \ldots + i_k = n$ as the numbers

$$C^n_{i_1, i_2, \ldots, i_k} = \frac{n!}{i_1! \, i_2! \cdots i_k!}$$

For the definition of the Fibonacci numbers and their various general-izations see Chapter I.

Problem 105

Let $\bar{x}_1, \bar{x}_2, \ldots, \bar{x}_n$ be non-negative integers. Prove that:

(a) In L_2 the number of *shortest* lattice paths connecting the point $(\bar{x}_1, \bar{x}_2)$ to the origin $0(0,0)$ is given by the binomial coefficient $C^{\bar{x}_1 + \bar{x}_2}_{\bar{x}_1}$.

(b) In L_3 the number of *shortest* lattice paths connecting the point $(\bar{x}_1, \bar{x}_2, \bar{x}_3)$ to $0(0,0,0)$ is given by the trinomial coefficient $C^{\bar{x}_1 + \bar{x}_2 + \bar{x}_3}_{\bar{x}_1, \bar{x}_2, \bar{x}_3}$.

(c) In general, in L_n the number of *shortest* lattice paths connecting the point $(\bar{x}_1, \bar{x}_2, \ldots, \bar{x}_n)$ to $0(0, 0, \ldots, 0)$ is given by the multinomial coefficient $C^{\bar{x}_1 + \bar{x}_2 + \ldots + \bar{x}_n}_{\bar{x}_1, \bar{x}_2, \ldots, \bar{x}_n}$.

From now on, in this chapter we shall consider only those lattice points in L_n whose coordinates are non-negative, that is, $\bar{x}_1, \bar{x}_2, \ldots, \bar{x}_n$ will stand for non-negative integers.

By labelling each point $(\bar{x}_1, \bar{x}_2, \ldots, \bar{x}_n)$ of L_n with the number $C^{\bar{x}_1 + \bar{x}_2 + \ldots + \bar{x}_n}_{\bar{x}_1, \bar{x}_2, \ldots, \bar{x}_n}$ of shortest lattice paths connecting it to O, we obtain an n-dimensional number pattern. In the two-dimensional case this pattern is the familiar Pascal's triangle (Fig. 1.7). The n-dimensional generalization of Pascal's triangle is often called the *n-dimensional Pascal pyramid*.

In the next problem the construction of the three-dimensional Pascal pyramid is carried out step by step, as follows.

Problem 106

In Σ_3 denote by λ_k the plane with equation

$$x_1 + x_2 + x_3 = k \quad \text{for} \quad i = 0, 1, 2, \ldots$$

λ_k intersects the coordinate axes $x_1, x_2,$ and x_3 in the points $A_k(k, 0, 0)$, $B_k(0, k, 0)$, and $C_k(0, k, 0)$ respectively (Fig. 1.45).

If $k = 0$, then $A_0 = B_0 = C_0 = (0, 0, 0)$, but if $k > 1$ then $A_k B_k C_k$ is an equilateral triangle.

(a) Make separate drawings of the triangles $A_k B_k C_k$ for $k = 1, 2, 3, 4$, marking in each triangle the lattice points $(\bar{x}_1, \bar{x}_2, \bar{x}_3)$ and labelling these points by their corresponding trinomial coefficients.

(b) Arrange the drawings in (a) underneath one another to obtain the initial part of the three-dimensional Pascal pyramid.

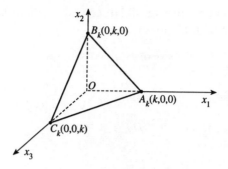

Fig. 1.45

In our Maths Club, the study of *two*-dimensional lattice paths, admitting diagonal steps, led us to patterns of points and of corresponding numbers in *three*-dimensional space.

We started with the following problem.

Problem 107

In L_2 define a lattice path, admitting diagonal steps, as follows (Fig. 1.46).

The path can contain (a) horizontal steps (from a point $(\bar{x}_1, \bar{x}_2)$ to $(\bar{x}_1 + 1, \bar{x}_2)$), (b) vertical steps (from a point $(\bar{x}_1, \bar{x}_2)$ to $(\bar{x}_1, \bar{x}_2 + 1)$), and (c) diagonal steps (along diagonals of cells, from $(\bar{x}_1, \bar{x}_2)$ to $(\bar{x}_1 + 1, \bar{x}_2 + 1)$).

(a) Prove that the number of lattice paths, connecting $O(0,0)$ with $\bar{x}_1\, \bar{x}_2$), which contain a given number r of diagonal steps is

$$\frac{(\bar{x}_1 + \bar{x}_2 - r)!}{(\bar{x}_1 - r)!\,(\bar{x}_2 - r)!\,r!}.$$

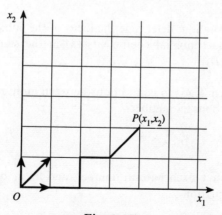

Fig. 1.46

(b) Deduce that the number of all lattice paths, admitting diagonal steps, which connect $O(0,0)$ with $(\bar{x}_1, \bar{x}_2)$ is

$$D_{\bar{x}_1, \bar{x}_2} = \sum_{r=0}^{\min (\bar{x}_1, \bar{x}_2)} \frac{(\bar{x}_1 + \bar{x}_2 - r)!}{(\bar{x}_1 - r)!(\bar{x}_2 - r)!r!}. \qquad (1.31)$$

(c) Denote by D_k the sum $\sum_{\bar{x}_1 = 0}^{k} D_{\bar{x}_1, k - \bar{x}_1}$; thus

$$D_k = \sum_{i=0}^{k} \sum_{r=0}^{\min (i, k-i)} \frac{(k - r)!}{r!(i - r)!(k - i - r)!}. \qquad (1.32)$$

Find a recurrence relation for D_k and hence show that D_k is the generalized Fibonacci number $f_n(2, 1)$, defined by (1.4) on p. 4.

We realized that the summands in (1.31) and in (1.32) were trinomial coefficients. Each trinomial coefficient is the label attached to some point in the cubic lattice L_3 in Σ_3. We decided to find the locus of the points in Σ_3 whose labels were the summands of $D_{\bar{x}_1, \bar{x}_2}$ and of D_k.

Problem 108

Let $\bar{x}_1, \bar{x}_2$ be given non-negative integers.
  Put

$$\left. \begin{array}{l} x_1 = \bar{x}_1 - r \\ x_2 = \bar{x}_2 - r \\ x_3 = r \end{array} \right\} \qquad (1.33)$$

Using the above notation, determine the locus of the cubic lattice points in Σ_3, labelled by the trinomial coefficients (a) in the sum for $D_{\bar{x}_1, \bar{x}_2}$, and (b) in the sum for D_k.

The next problem leads to connections between multinomial coefficients and generalized Fibonacci numbers.

Problem 109

Let k be an arbitrary fixed natural number, and let π_k be the plane in Σ_3 (Fig. 1.47) with equation

$$x_1 + x_2 + 2x_3 = k.$$

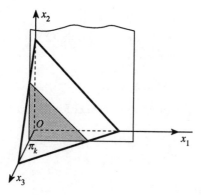

Fig. 1.47

(a) By intersecting π_k with planes, perpendicular to the Ox_3-axis, prove that

$$D_k = \sum_{r=0}^{\left[\frac{k}{2}\right]} \binom{k-r}{r} 2^{k-2r} \tag{1.34}$$

(b) Prove that the sum of the trinomial coefficients attached to the lattice points $(\bar{x}_1, i, \bar{x}_3)$ of π_k, with a fixed second coordinate $i \in \{0, 1, \ldots, k\}$, is a Fibonacci number of order i (defined in Chapter I, Section (1.4)).

(c) Hence, express D_k as a sum of Fibonacci numbers of higher order.

By equating pairwise the expressions for D_k, obtained in Problem 107 (c) and in Problem 109 (a) and (c), one obtains connections between trinomial and binomial coefficients, as well as between generalized Fibonacci numbers and trinomial or binomial coefficients. What makes these connections especially interesting is that they can be viewed as extensions of a famous formula for Fibonacci numbers due to the nineteenth-century mathematician Édouard Lucas.

Lucas proved that the kth Fibonacci number f_k is the sum of the binomial coefficients $\binom{k-r}{r}$ for $r = 0, 1, \ldots, \left[\frac{k}{2}\right]$, that is

$$f_k = \sum_{r=0}^{\left[\frac{k}{2}\right]} \binom{k-r}{r}. \tag{1.35}$$

We are now able to relate Lucas' result to the results of Problem 109:

Problem 110

Show that (1.35) is a special case of formula (2.92) obtained in part (b) of Problem 109.

Problem 111

For an arbitrary fixed natural number n consider the sum

$$\rho_{k,n} = \sum_{r=0}^{\left[\frac{n}{2}\right]} \binom{k-r}{r} n^{k-2r},$$

generalizing (1.35) and (1.34).

(a) Find a recurrence relation for $\rho_{k,n}$ and show that $\rho_{k,n}$ is a generalized Fibonacci number.

(b) Solve the recurrence relation for $\rho_{k,n}$ (see Chapter I, Section 1.5) to obtain a generalization of Binet's formula (1.6).

Finally, an application of lattice paths in the study of the *problem of the gambler's ruin* (well known in the theory of probability, where one asks for the chances of the gambler being ruined).

Suppose a gambler has a fortune of £p (p a natural number) and makes n ($n \geqslant p$) successive bets of £1 each. If bankruptcy occurs, he (she) is permitted to complete the run of n bets, but the run is considered ruinous. Our aim here is to find the number of ruinous sequences for the gambler.

To do this we shall paraphrase the problem geometrically, by using a square lattice (Fig. 1.48).

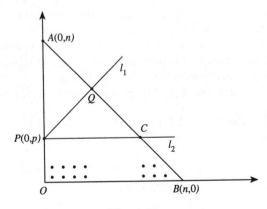

Fig. 1.48

In this lattice each run of n bets will be represented by a lattice path of n steps starting from the origin O, so that each gain of £1 corresponds to a horizontal step to the right ($\rightarrow$), and each loss of £1 to a vertical step up ($\uparrow$).

The lattice in Fig. 1.48 will be used for counting the number of ruinous sequences.

Problem 112

In Fig. 1.48 plot the points $A(0, n)$, $B(n, 0)$, $P(0, p)$ and draw the line segment AB and the lines ℓ_1 with equation $y = x + p$, and ℓ_2 with equation $y = p$. Denote by Q and C the intersections of AB with ℓ_1 and ℓ_2 respectively.

(a) Verify that each ruinous sequence of bets is represented by one of the shortest paths starting at O, ending at AC, *and* having at least one point in common with PQ.

(b) Count the number of lattice paths described in (a), and hence find the number of sequences of n bets leading to the gambler's ruin.

IX Do it with groups

Introduction

The theory of groups is an important branch of modern algebra, with various applications in mathematics and in physics. The notion of a *group* is based on that of a *binary operation on a set*.

A *binary operation* on a non-empty set S is a rule which associates to every ordered pair (a, b) of (not necessarily distinct) elements a, b of S a unique element of S.

For example, addition is a binary operation on the set $\mathbb{N}$ of natural numbers; it associates to each ordered pair (a, b) of natural numbers their sum $a + b \in \mathbb{N}$.

A binary operation on S is usually denoted by a symbol, say '$\circ$', and in that case the element of S associated by $\circ$ to (a, b) is denoted by $a \circ b$.

A *group* is a non-empty set of elements, G, with a binary operation $\circ$ defined in G, which satisfies the following conditions:

G_1. The operation $\circ$ is *associative*, that is, for every triple (a, b, c) of elements in G, $(a \circ b) \circ c = a \circ (b \circ c)$.

G_2. There is a unique element $e \in G$, called the identity, such that $e \circ a = a \circ e = a$ for every $a \in G$.

G_3. Each element $a \in G$ has an inverse $a^* \in G$ such that $a \circ a^* = a^* \circ a = e$.

Conditions G_1, G_2, and G_3 are called the *group axioms*.

There are many familiar examples of groups, for example

the set $\mathbb{Z}$ of all integers under addition (in this group the identity is 0, and the inverse of each $a \in \mathbb{Z}$ is $-a$),

the set $Q \setminus \{0\}$ of all non-zero rational numbers under multiplication with identity 1 and inverse a^{-1} for each $a \in Q$,

the set $\{1, -1\}$ under multiplication, and

the set S of all rotations of the plane about a given point under composition of rotations (for $a, b \in S$ the composition $a \circ b$ of the rotations a, b means first perform rotation b and then rotation a).

The aim of this chapter is to point out that groups arise in connection with various problems, and that their properties can be used to solve the problems in question. We shall use groups to solve some functional equations (in Section 9.2), to study properties of lattice points in the plane (in Section 9.3), and to discuss solutions of puzzles involving permutations (in Section 9.4).

Section 9.1 is preliminary; it gives examples and properties of groups, needed in the rest of the chapter.

9.1   Examples and some basic properties of groups

Problem 113

Let n be a given natural number, and let $\mathbb{Z}_n = \{0, 1, 2, \ldots, n-1\}$ be the set of their remainders upon division by n. Define two binary operations on $\mathbb{Z}_n$:

(1) *addition modulo n*, denoted by $\oplus$, where $a \oplus b$ is the remainder of $a + b$ when divided by n for all $a, b \in \mathbb{Z}_n$;

(2) *multiplication modulo n*, denoted by $\otimes$, where $a \otimes b$ is the remainder of $a \cdot b$ when divided by n for all $a, b \in \mathbb{Z}_n$.

(a) Prove that $\mathbb{Z}_n$ is a group with respect to addition modulo n for every natural number n.

(b) Give a necessary and sufficient condition for n under which the set $\mathbb{Z}_n^* = \{1, 2, 3, \ldots, n-1\}$ forms a group with respect to multiplication modulo n.

Problem 114

Consider the three functions, defined on the set

$S = \{x : x \text{ is a rational number}, x \neq 0, x \neq 1\}$:

$$f_1(x) = x, \quad f_2(x) = \frac{1}{1-x}, \quad f_3(x) = \frac{x-1}{x}:$$

Prove that $\{f_1(x), f_2(x), f_3(x)\}$ forms a group under the operation of composition f functions ($\circ$).

Remark. Recall that the composition $\circ$ of two functions $f_i(x)$ and $f_j(x)$ is defined as

$$f_i(x) \circ f_j(x) = f_i(f_j(x)) \quad \text{for all } x \in S.$$

Problem 115

Denote by Q' the union of the set Q of rational numbers and of the set $\{\infty\}$.
For each ordered quadruple (a, b, c, d) of integers such that $ad - bc = 1$,
define the function $f_{a,b,c,d}(x)$ on Q' as follows.

(a) If $c \neq 0$ then

$$f_{a,b,c,d}(x) = \begin{cases} \dfrac{ax+b}{cx+d} & \text{if } x \neq -\dfrac{d}{c} \text{ and } x \neq \infty \\[2mm] \dfrac{a}{c} & \text{if } x = \infty \\[2mm] \infty & \text{if } x = -\dfrac{d}{c}. \end{cases}$$

(b) If $c = 0$ then

$$f_{a,b,c,d}(x) = \begin{cases} \dfrac{ax+b}{d} & \text{if } x \neq \infty \\[2mm] \infty & \text{if } x = \infty \end{cases}$$

Prove that these functions form a group under the operation of composition
of functions.

Now we turn to an important class of algebraic structures known as
permutation groups.

A *permutation* π of a set S is a one-to-one mapping of S onto itself. π is
often denoted by a rectangular array of two rows: the upper row consists
of the elements of S and the lower row of their images under π:

$$\pi = \begin{pmatrix} \cdot & \cdot & \cdot & \cdot & \cdot & x & \cdot & \cdot & \cdot \\ \cdot & \cdot & \cdot & \cdot & \cdot & \pi(x) & \cdot & \cdot & \cdot \end{pmatrix}$$

Problem 116

(a) Prove that all permutations on a set S form a group under the
operation of composition $\circ$. (For any two permutations π, π' on S and
for any $x \in S$, the image of x under $\pi \circ \pi'$ is $\pi \circ \pi'(x) = \pi(\pi'(x))$.) This
group is called the *full symmetric group* on S.

(b) Show that if S has n elements, then the full symmetric group on S has
$n!$ elements.

Our next step is to define the *sign* of a permutation on a *finite set* S (that is,
on a set S with finitely many elements), as follows.

Consider a permutation

$$\pi = \begin{pmatrix} a_1 & a_2 & \cdot & \cdot & \cdot & \cdot & a_n \\ \pi(a_1) & \pi(a_2) & \cdot & \cdot & \cdot & \cdot & \pi(a_n) \end{pmatrix}.$$

Whenever two symbols a_i, a_j appear in a certain order in the top row and in the reverse order in the bottom row, then we say that π has an *inversion*.

For example, the permutation $\begin{pmatrix} 1 2 3 4 5 \\ 3 4 2 1 5 \end{pmatrix}$ has five inversions, each of them represented in Fig. 1.49 by a pair of intersecting lines drawn between pairs of identical symbols.

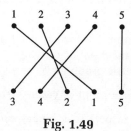

Fig. 1.49

The *sign* of a permutation π of a finite set S is defined as

$$\text{sign } \pi = (-1)^{v}$$

where v is the number of inversions of π.

Problem 117

Prove the following properties of signs of permutations.

(a)   If $\pi = \begin{pmatrix} a_1 & a_2 & \cdot & \cdot & \cdot & \cdot & a_n \\ \pi(a_1) & \pi(a_2) & \cdot & \cdot & \cdot & \cdot & \pi(a_n) \end{pmatrix}$ then

$$\text{sign } \pi = \frac{(a_1 - a_2)(a_1 - a_3)\ldots(a_1 - a_n)}{[\pi(a_1) - \pi(a_2)][\pi(a_1) - \pi(a_3)]\ldots[\pi(a_1) - \pi(a_n)]}$$

$$\cdot \frac{(a_2 - a_3)\ldots(a_2 - a_n)}{[\pi(a_2) - \pi(a_3)]\ldots[\pi(a_2) - \pi(a_n)]} \cdot \ldots \cdot \frac{a_{n-1} - a_n}{\pi(a_{n-1}) - \pi(a_n)}$$

(b) Deduce that if π and π' are two permutations on a set S then

$$\text{sign } (\pi \circ \pi') = \text{sign } \pi \cdot \text{sign } \pi'.$$

A permutation with an even number of inversions is called an *even permutation*, and a permutation with an odd number of inversions is called an *odd permutation*.

Problem 118

(a) Prove that the even permutations of a set S of n elements form a group with respect to the operation of composition $\circ$. This group is called the *alternating group* on S.

(b) Prove that the alternating group on S has $\frac{1}{2}n!$ elements. Thus the number of even permutations on S is equal to the number of odd permutations on S.

Remark. A subset of a group G which is itself a group under the group operation on G is called a *subgroup* of G. Thus the alternating group on S is a subgroup of the symmetric group on S.

Finally, we introduce the concept of *generators* of a group.

Let G be a group with respect to a binary operation $\circ$. We say that a subset H of G *generates* G if every element of G can be written as a 'product' $h_1 \circ h_2 \circ h_3 \circ \ldots \circ h_k$, where k is a natural number, and $h_1, h_2, \ldots, h_k$ are k (not necessarily distinct) elements of H. The elements of H are called *generators* of G.

For example, the set $\{1, -1\}$ generates the group $\mathbb{Z}(+)$ of all integers with respect to addition because

any positive integer n is equal to $\underbrace{1 + 1 + \ldots + 1}_{n}$,

any negative integer $-m$ is equal to $\underbrace{-1 + (-1) + \ldots + (-1)}_{m}$

and $0 = 1 + (-1)$.

Problem 119

(a) Prove that the symmetric group σ_3 on the set $S = \{1, 2, 3\}$ can be generated by two permutations of S. (Show that there are more pairs of generators for σ_3.)

(b) What is the smallest number of generators of the group $\mathbb{Z}_n(\oplus)$, defined in Problem 113?

We are now able to look at some applications of groups.

9.2   How to use groups for the solution of some functional equations

Functional equations are equations in which the unknowns are functions. To solve functional equations is to determine the unknown functions in these equations.

The study of functional equations, prompted by problems in mechanics, began more than 200 years ago. There is a great variety of functional equations and the methods for solving them differ greatly. In this section we shall consider some functional equations whose solution can be facilitated by the use of elementary properties of groups. This is illustrated by our solution of Problems A and B below.

Problem A. Find all functions $f(x)$ such that

$$xf(x) = 1 + 2f(1 - x) \text{ for all real numbers } x. \tag{1.36}$$

Solution. Suppose that there is a function f satisfying (1.36). In (1.36) the function f is applied to x and $1 - x$. This suggests the idea of replacing x by $1 - x$ in (1.36) and of rewriting (1.36) accordingly:

$$(1 - x)f(1 - x) = 1 + 2f[1 - (1 - x)],$$

that is,

$$(1 - x)f(1 - x) = 1 + 2f(x). \tag{1.37}$$

We are 'lucky': in (1.37), as in (1.36), the function f is applied to x and $1 - x$. Hence (1.36) and (1.37) can be considered as a system of simultaneous linear equations with unknowns $f(x)$ and $f(1 - x)$. By solving this system we find that

$$f(x) = \frac{x - 3}{x^2 - x + 4}. \tag{1.38}$$

(Verify this!)

Since $x^2 - x + 4 > 0$ for all real numbers x, the expression in (1.38) is defined for all real numbers x. It remains to verify that (1.38) satisfies the functional equation (1.36). This is done by substituting

$$f(x) = \frac{x - 3}{x^2 - x + 4} \text{ and } f(1 - x) = \frac{(1 - x) - 3}{(1 - x)^2 - (1 - x) + 4}$$

in (1.36). This finishes the solution of Problem A.

Why was the above method successful? This can be explained by using the concept of a group:

The change of x to $1 - x$ corresponds to the function $g_2(x) = 1 - x$, while keeping x unchanged corresponds to the function $g_1(x) = x$.

It is easy to verify that the set $\{g_1(x), g_2(x)\}$ forms a group with respect to the operation of composition of functions ($\circ$). Therefore we can rephrase the steps in the solution of (1.36) as follows.

Step 1. Rewrite (1.36) in terms of $g_1(x)$ and $g_2(x)$:

$$xf(g_1(x)) = 1 + 2f(g_2(x)). \tag{1.39}$$

Step 2. Replace the variable x in (1.39) by $g_2(x)$; this yields

$$g_2(x)f(g_1(g_2(x))) = 1 + 2f(g_2(g_2(x))),$$

that is,

$$(1 - x)f(\underbrace{g_1(x) \circ g_2(x)}) = 1 + 2f(\underbrace{g_2(x) \circ g_2(x)})$$

or

$$(1 - x)f(g_2(x)) \qquad = 1 + 2f(g_1(x)) \tag{1.40}$$

The system of simultaneous equations (1.39) and (1.40) with unknowns $f(g_1(x))$ and $f(g_2(x))$ is essentially the same as the system consisting of (1.36) and (1.37); thus it can be solved in the same way, leading to

$$f(g_1(x)) = f(x) = \frac{x - 3}{x^2 - x + 4}.$$

Problem B. Solve the functional equation

$$xf(x) + 2f\left(\frac{x - 1}{x + 1}\right) = 1. \tag{1.41}$$

Solution.

Step 1. Suppose that (1.41) has a solution. Put

$$g_1(x) = x \quad \text{and} \quad g_2(x) = \frac{x - 1}{x + 1}.$$

Find all elements of the group $G(\circ)$, generated by $g_1(x)$ and $g_2(x)$ under the operation $\circ$ of composition of functions.

It is easy to verify that $G(\circ)$ has four elements:

$$g_1(x), g_2(x), g_3(x) = -\frac{1}{x}, \quad \text{and} \quad g_4(x) = \frac{x + 1}{1 - x}.$$

The functions $g_1(x), i = 1, 2, 3, 4$, are defined for all real numbers x such that $x \neq -1, 0, 1$.

Step 2. Rewrite (1.41) in the form

$$xf(g_1(x)) + 2f(g_2(x)) = 1 \tag{1.42}$$

and replace x in (1.42) by $g_i(x)$ for $i = 2, 3, 4$. This leads to three new equations:

$$\frac{x-1}{x+1} f(g_2(x)) + 2f(g_3(x)) = 1, \tag{1.43}$$

$$-\frac{1}{x} f(g_3(x)) + 2f(g_4(x)) = 1, \tag{1.44}$$

$$\frac{x+1}{1-x} f(g_4(x)) + 2f(g_1(x)) = 1. \tag{1.45}$$

Step 3. (1.42)–(1.45) form a system of four simultaneous equations with unknowns $f(g_i(x))$, $i = 1, 2, 3, 4$. By solving this system one obtains $f(x)$ (that is $f(g_1(x))$) in the form

$$f(x) = \frac{4x^2 - x + 1}{5x(x-1)} \tag{1.46}$$

Step 4. By substituting $f(x) = \dfrac{4x^2 - x + 1}{5x(x-1)}$ and

$$f\left(\frac{x-1}{x+1}\right) = \left[4\left(\frac{x-1}{x+1}\right)^2 - \frac{x-1}{x+1} + 1\right] \Big/ \left[5\left(\frac{x-1}{x+1}\right)\left(\frac{x-1}{x+1} - 1\right)\right]$$

in (1.41) verify that (1.46) is a solution of (1.42) for all real numbers x such that $x \notin \{-1, 0, 1\}$.

The reader should try to solve the next two problems.

Problem 120

Solve the functional equation

$$xf(x) + 2f\left(-\frac{1}{x}\right) = 3 \text{ for real numbers } x \neq 0.$$

Problem 121

Solve the functional equation

$$af(x^n) + f(-x^n) = bx \text{ for real numbers } x$$

where a, b, and n are given numbers such that $a \neq 1$ and n is an odd integer.

9.3 Groups and lattice points

In a two-dimensional rectangular cartesian coordinate system each point $P(x, y)$ has a corresponding *position vector* $\mathbf{OP} = \begin{pmatrix} x \\ y \end{pmatrix}$ with respect to the origin $O(0, 0)$; this is the vector with initial point O and endpoint P (Fig. 1.50(a)).

The *sum* of two position vectors $\mathbf{OP} = \begin{pmatrix} x \\ y \end{pmatrix}$ and $\mathbf{OP'} = \begin{pmatrix} x' \\ y' \end{pmatrix}$ is defined as the position vector $\mathbf{OP''} = \begin{pmatrix} x + x' \\ y + y' \end{pmatrix}$.

The above definition has a geometrical interpretation: the endpoint of $\mathbf{OP''}$ is the fourth vertex of the parallelogram with sides OP and OP'.

Problem 122

Prove the following.

(a) The set V of vectors $\begin{pmatrix} x \\ y \end{pmatrix}$, where x and y arbitrary real numbers, forms a group $V(+)$ with respect to the operation of vector addition.

(b) The set H of vectors $\begin{pmatrix} m \\ n \end{pmatrix}$, where m and n are arbitrary integers, forms a subgroups of $V(+)$.

The endpoints of the vectors $\begin{pmatrix} m \\ n \end{pmatrix}$ of H are the lattice points of the square lattice defined in Chapter III, Section 3.3 (Fig. 1.50(b). *Our aim is to show that properties of the group $H(+)$ can be applied in the study of 'lattice parallelograms', that is, parallelograms whose vertices are lattice points (see Problem 1.25).*

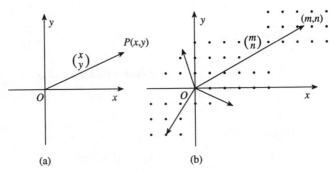

(a)   (b)

Fig. 1.50

First one must solve the following two problems.

Problem 123

(a) Verify that the group $H(+)$ can be generated by the two vectors $\boldsymbol{OI} = \begin{pmatrix} 1 \\ 0 \end{pmatrix}$ and $\boldsymbol{OJ} = \begin{pmatrix} 0 \\ 1 \end{pmatrix}$.

(b) Prove that two vectors $\begin{pmatrix} a \\ b \end{pmatrix}$ and $\begin{pmatrix} c \\ d \end{pmatrix}$ of $H(+)$ generate $H(+)$ if and only if

$$|ad - bc| = 1. \tag{1.47}$$

The number $|ad - bc|$ has a geometrical meaning:

Problem 124

Prove that $|ad - bc|$ is the area of the parallelogram with vertices $O((0,0)$, $P(a,b)$, $R(a+c,b+d)$, and $Q(c,d)$.

Combining the statements in Problems 123 and 124 we are able to prove the property (ℙ) of lattice parallelograms, formulated in Chapter III, Section 3.3:

Problem 125

Let P be a parallelogram whose vertices are lattice points. Prove that if there are no other lattice points on the perimeter of P or inside P, then the area of P is 1.

9.4 Permutations in the study of the fifteen puzzle and Rubik's cube

In this section we shall consider two famous games: the 'fifteen puzzle', which was very popular at the end of the last century, and Rubik's cube, invented in the 1970s. We shall study *necessary* conditions for the solvability of these games.

The *fifteen puzzle* consists of fifteen counters and a container box. The counters are congruent cuboids of dimensions a, a, b and the container is a cuboid of dimensions $4a, 4a$, and b. The counters are labelled 1 to 15.

At the start of the game the counters are placed in the container in an arbitrary way, leaving the bottom right corner of the container empty (Fig. 1.51). The object of the puzzle is to shift the counters in the container parallel to the edges of the container's base until their labels are arranged in an increasing order, as shown in Fig. 1.51.

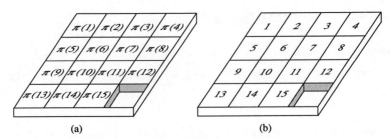

Fig. 1.51

Let us call the arrangement of the counters with labels in increasing order their *basic arrangement*. An arbitrary arrangement of the counters, leaving the bottom right corner of the container empty, represents a permutation π of the basic arrangement:

$$\pi = \begin{pmatrix} 1 & 2 & 3 & \cdots\cdots & 15 & 16 \\ \pi(1) & \pi(2) & \pi(3) & & \pi(15) & \cdot \end{pmatrix}. \tag{1.48}$$

Since 15 objects can be permuted in 15! different ways, it follows that the counters can be arranged in the box in 15! different ways at the start of the game. The question is

Can the puzzle be solved, starting with any of the 15! arrangements of the counters?

The statement of the next problem provides a negative answer to the question.

Problem 126

Prove that a necessary condition for the solvability of the fifteen puzzle is that the permutation π, corresponding to the initial arrangement of the counters in the container, is even.

Remark. It can be proved that the condition stated in Problem 125 is also sufficient for the solvability of the puzzle. The proof of this is not difficult, but rather lengthy, and therefore will be omitted. Readers may test their skills by trying to solve the puzzle with initial arrangements corresponding to various even permutations of the labels on the counters.

Rubik's cube was invented by the Hungarian architect E. Rubik. It consists of 26 pieces, called *cubelets*, assembled around a 'core' to form a cube (Fig. 1.52(a)). Each face of C is constructed from nine small squares, which are faces of nine cubelets, forming a *layer* of C. Any layer can be rotated about the axis connecting its centre to the centre of C. Obviously, a rotation of a

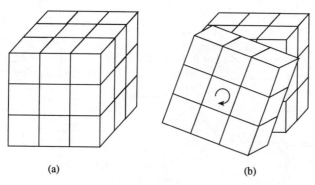

(a)   (b)

Fig. 1.52

layer through an angle which is not a multiple of 360° destroys the layers on the adjacent faces of C. On the other hand, each rotation of any layer through a multiple of 90° maps the cube C onto itself (Fig. 1.52).

At a certain 'initial' stage the faces of C are painted in six different colours. After that layers are turned through angles of $k \cdot 90°$ (for $k \in \{0, \pm1, \pm2, \ldots\}$ any number of times in an arbitrary order. This procedure may transform C so that its faces contain various patterns of differently coloured small squares.

The aim of the puzzle is to start with a given arbitrary arrangement of colours on the faces of C and to restore this arrangement to the initial stage (in which all faces of C are monochromatic) by rotating the layers.

When Rubik's cube appeared on the market, it attracted a great deal of attention not only among puzzle fans but also among mathematicians; methods for 'solving' the cube were published, accompanied by explanations of their mathematical background.

In this section we shall study aspects of an interesting question, described by Dubrovski [35].

The mechanism of Rubik's cube is such that the eight cubelets at the vertices of C (the 'vertex cubelets') and the 12 cubelets in the middle of C's edges (the 'edge cubelets') can be removed individually from the cube, while the six cubelets at the centre of the faces (the 'central cubelets') are firmly attached to the core (Fig. 1.53).
Dubrovski asks:

Suppose that the corner cubelets and the edge cubelets are removed from C, and then C is reconstructed by inserting these pieces in the slots around the core in an arbitrary way. Is it possible to 'solve the cube' starting from any such arrangement of the cubelets?

We shall show (following Dubrovski) that the answer to this question is 'No', by giving *necessary* conditions for the solvability of the reassembled cube.

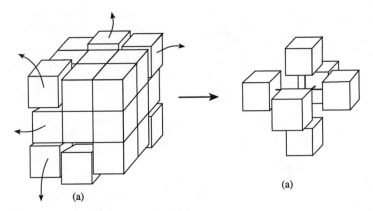

Fig. 1.53

To do this, consider first the initial arrangement A of the cubelets around the core of C in which each face of C shows only one colour. Suppose that the top face of C is blue, and the opposite face is green. Label the vertex cubelets 1 to 8, and the edge cubelets 1 to 12. We can imagine that the removable cubelets are fitted into empty cells when C is reassembled. Number these cells using the labels of the cubelets which they hold in the initial arrangement A (Fig. 1.54).

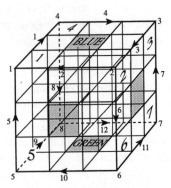

Fig. 1.54

The above labelling allows one to assign to each arrangement Π of the cubelets in the cells of C two permutations:

(1) the permutation

$$\sigma_\Pi = \begin{pmatrix} 1 & 2 & 3 & 4 & 5 & 6 & 7 & 8 \\ \sigma_\Pi(1) & \sigma_\Pi(2) & \sigma_\Pi(3) & \sigma_\Pi(4) & \sigma_\Pi(5) & \sigma_\Pi(6) & \sigma_\Pi(7) & \sigma_\Pi(8) \end{pmatrix},$$

showing the arrangement of the vertex cubelets, $\sigma_\Pi(i)$ is the label of the vertex cubelet occupying the ith corner cell for $i = 1, \ldots, 8$, and

(2) the permutation

$$\tau_\Pi = \begin{pmatrix} 1 & 2 & 3 & 4 & 5 & 6 & 7 & 8 & 9 & 10 & 11 & 12 \\ \tau_\Pi(1) & \tau_\Pi(2) & . & . & . & . & . & . & . & . & . & \tau_\Pi(12) \end{pmatrix}$$

showing the arrangement of the edge cubelets, $\tau_\Pi(i)$ is the label of the edge cubelet in the ith edge cell for $i = 1, \ldots, 12$.

With the help of the above definitions we can formulate the first necessary condition for the solution of the cube:

Problem 127

Let Π be a given arrangement of the cubelets around the core of C, with corresponding permutations σ_Π and τ_Π. Prove that a necessary condition for the solvability of the cube, starting from Π, is that

$$\text{sign } \sigma_\Pi \cdot \text{sign } \tau_\Pi = 1 \qquad (1.49)$$

(that is, either both permutations σ_Π and τ_Π are even, or both are odd).

Condition (1.49) does not seem to be sufficient for solving the cube, since a loose cubelet can be fitted into the same cell with its painted faces in different positions. This is indeed the case. We shall formulate two further conditions necessary for restoring C to its elementary stage:

One of these conditions concerns the fitting of the vertex cubelets. Each such cubelet has either a blue face or a green face. For each $i = 1, \ldots, 8$ consider the cubelet $\sigma_\Pi(i)$ in the ith corner cell. Denote by $\alpha_\Pi(i)$ the angle through which the cubelet should be turned anticlockwise about the diagonal of C so that its blue or green face appears on the face of C in which the central cubelet has one of these two colours — blue or green. Technically such a turn is impossible without removing $\sigma_\Pi(i)$ from C. Hence

Problem 128

(a) Prove that the sum

$$\Sigma'_\Pi \equiv \alpha_\Pi(1) + \alpha_\Pi(2) + \ldots + \alpha_\Pi(8) \pmod{360°}$$

is invariant under the rotations of the layers which map C onto itself.

(b) Deduce that the condition

$$\Sigma'_\Pi \equiv 0 \pmod{360°} \qquad (1.50)$$

is necessary for any initial arrangement of Π of the cubelets which leads to a 'solution' of the cube.

The next condition involves the edge cubelets. The study of their positions is simplified by the following procedure.

In the basic arrangement A draw an arrow along that edge of the ith edge cubelet which belongs to an edge of C. The arrow can be drawn in either of the two possible directions, but once drawn it remains fixed. Moreover, affix an arrow in the same direction to the corresponding edge of the empty cell, which in arrangement A holds the ith edge cubelet. Do this for $i = 1, 2, \ldots, 12$.

In an arbitrary arrangement Π of the cubelets, the edge cubelet $\tau(i)$ in the ith cell has its arrow either in the same direction as the arrow of the ith cell, or in the opposite direction. Denote the angle between these two arrows by $\beta_\Pi(i)$.

Problem 129

(a) Prove that the sum

$$\textstyle\sum_{\Pi}'' = \beta_\Pi(1) + \beta_\Pi(2) + \ldots + \beta_\Pi(12) \pmod{360°}$$

is invariant under the rotations of the layers which map C onto itself.

(b) Deduce that the condition

$$\textstyle\sum_{\Pi}'' \equiv 0 \pmod{360°} \tag{1.51}$$

is necessary for any initial arrangement Π of the cubelets which leads to a 'solution' of the cube.

In view of conditions (1.49), (1.50), and (1.51) we can restrict the number of arrangements of the cubelets leading to the 'solution' of the cube.

Problem 130

(a) Prove that the loose vertex cubelets and edge cubelets can be assembled around the core to form the cube C in

$$8! \cdot 3^8 \cdot 12! \cdot 2^{12}$$

different ways.

(b) Show that in view of (1.49), (1.50), and (1.51), at least $8! 3^8 \cdot 10! 11^2 \cdot 2^{12}$ of the possible arrangements do not lead to the solution of the cube.

Remark. Dubrovski shows in his article that conditions (1.49), (1.50), and (1.51) are also *sufficient* for the solution of Rubik's puzzle. We shall omit the (rather lengthy) proof of this result.

X From puzzles to research topics: selected problems of combinatorics

Introduction

Various puzzles prompted important research in combinatorics and related fields of mathematics. *Euler's 36 officers* problem is a famous example:

36 officers of 6 ranks and from 6 regiments have to be arranged in a 6×6 square formation so that every row and every column of the formation includes officers of each rank and each regiment.

Euler's study of the above puzzle initiated the study of *Latin squares* which nowadays play an important role in statistics and in various branches of discrete mathematics. Our aim in the first part of this chapter is to explain the connection between the puzzle and Latin squares, to solve problems on Latin rectangles and Latin squares (in Section 10.1), and to point out an important link between Latin squares and a class of non-Euclidean geometries — the finite projective planes (in Section 10.2).

The second part of Chapter X is devoted to problems on map colouring. Section 10.3 introduces the reader to the celebrated four colour conjecture:

Four colours are sufficient to colour any geographical map subject to the restriction that neighbouring countries have different colours.

The last part of the chapter (Section 10.4) is concerned with problems which can be solved by using graphs.

10.1 36 officers and Latin squares

A *Latin square of order n* is an $n \times n$ square array of n symbols such that each row and each column of the array contains every symbol.

Below are three Latin squares of order 4 with entries 1, 2, 3, and 4:

$$
M = \begin{pmatrix} 1 & 2 & 3 & 4 \\ 2 & 1 & 4 & 3 \\ 3 & 4 & 1 & 2 \\ 4 & 3 & 2 & 1 \end{pmatrix} \quad N = \begin{pmatrix} 1 & 2 & 3 & 4 \\ 4 & 3 & 2 & 1 \\ 2 & 1 & 4 & 3 \\ 3 & 4 & 1 & 2 \end{pmatrix} \quad K = \begin{pmatrix} 1 & 2 & 3 & 4 \\ 2 & 3 & 4 & 1 \\ 3 & 4 & 1 & 2 \\ 4 & 1 & 2 & 3 \end{pmatrix}.
$$

The problem of the 36 officers leads to two Latin squares of order 6 as follows.

Suppose that the 36 officers are arranged in a 6×6 square formation of the required kind. Denote the rank of the officer in the ith row and jth column by r_{ij} and his regiment by t_{ij}. Label the ranks and the regiments 1, 2, 3, 4, 5, 6. Thus $r_{ij}, t_{ij} \in \{1, 2, 3, 4, 5, 6\}$. Attach to each officer in the formation the ordered pair of numbers (r_{ij}, t_{ij}).

Since in the formation each rank must appear in every row and in every column exactly once, the matrix $R = (r_{ij})$ is a Latin square of order 6. Similarly, since each regiment must appear in every row and in every column exactly once, the matrix $T = (t_{ij})$ is also a Latin square of order 6.

The two Latin squares R and T must satisfy an additional condition: the ordered pairs (r_{ij}, t_{ij}) of labels must be different for all $i, j \in \{1, 2, 3, 4, 5, 6\}$ (otherwise the same officer would stand in two different places). The Latin squares R and T satisfying this additional condition are called *orthogonal*.

In general, we have the definition.

Let $A = (a_{ij})$ and $B = (b_{ij})$ be two Latin squares of order n with $a_{ij}, b_{ij} \in \{1, 2, \ldots, n\}$. The Latin squares A and B are called *orthogonal* provided the ordered pairs of numbers (a_{ij}, b_{ij}) are distinct for $i, j = 1, 2, \ldots, n$.

(From the three Latin squares of order 4 given above, M and N form an orthogonal pair, while M and K, and N and K are not orthogonal.)

Euler attempted to construct two orthogonal Latin squares of order 6 but did not succeed. He suspected that this is impossible. In 1782 he formulated the following general conjecture:

There exists no pair of orthogonal Latin squares of order n for any $n = 4k + 2$, where $k = 0, 1, 2, \ldots$.

Around 1900 G. Tarry verified the non-existence of orthogonal Latin squares of order 6. (Thus the 36 officers cannot be arranged in a square formation satisfying the conditions of the puzzle.) Sixty years later Euler's conjecture was disproved by the combined efforts of Bosé, Shrikhande, and Parker, who established the following theorem.

Theorem T_1. If $m = 4k + 2$, where $k = 2, 3, 4, \ldots$, then there exists a pair of orthogonal Latin squares of order n.

This famous example illustrates the danger of jumping to general conclusions on the basis of a few instances which indicate a pattern.

Theorem T_1 has a counterpart, due to MacNeish (around 1922).

Theorem T_2. If n is a natural number, greater than 1, and not of the form $4k + 2$ for $k = 0, 1, 2, \ldots$, then there exists a pair of orthogonal Latin squares of order n.

T_1 and T_2 imply that for all natural numbers n except 1, 2, and 6 one can construct pairs of orthogonal Latin squares of order n.

The proofs of theorems T_1 and T_2 are lengthy and will be omitted. Instead, we list a number of problems on Latin rectangles and Latin squares, pointing out connections with other mathematical topics.

Latin rectangles are generalizations of Latin squares:
An r by n Latin rectangle with entries $1, 2, \ldots, n$ *is an* $r \times n$ *rectangular array in which each row is a permutation of the set* $\{1, 2, \ldots, n\}$, *and these permutations are chosen so that each column consists of distinct elements.*

For example,

$$P = \begin{pmatrix} 2 & 3 & 1 & 4 \\ 1 & 4 & 2 & 3 \end{pmatrix} \text{ is a } 2 \times 4 \text{ Latin rectangle with entries } 1, 2, 3, 4.$$

An r by n Latin rectangle with entries $1, 2, \ldots, n$ *is called normalized if its first row consists of the numbers* $1, 2, \ldots, n$ *arranged in their natural, increasing order.*

(For example, $Q = \begin{pmatrix} 1 & 2 & 3 & 4 \\ 3 & 1 & 4 & 2 \\ 4 & 2 & 1 & 3 \end{pmatrix}$ is a normalized 3×4 Latin rectangle.)

Combinatorists became interested in the problem of finding the number of distinct normalized $r \times n$ Latin rectangles with entries $1, 2, \ldots, n$. This is still an open problem. Some special cases have yielded interesting answers.

Problem 131

Prove that the number of the normalized $2 \times n$ Latin rectangles with entries $1, 2, \ldots, n$ is

$$D_n = n! \left[1 - \frac{1}{1!} + \frac{1}{2!} - \ldots + (-1)^n \frac{1}{n!} \right] \text{ for all } n \geqslant 2.$$

Remark. A normalized $2 \times n$ Latin rectangle

$$\begin{pmatrix} 1 & 2 & \ldots & n \\ a_{21} & a_{22} & \ldots & a_{2n} \end{pmatrix}$$

corresponds to a permutation π of $\{1, 2, \ldots, n\}$ with $\pi(i) = a_{2i} \neq i$ for all $i = 1, 2, \ldots, n$. A permutation in which no element is mapped onto itself is called a *derangement*. Thus D_n is the number of derangements of $\{1, 2, \ldots, n\}$.

The enumeration of a special class of normalized $3 \times n$ Latin rectangles leads to the celebrated 'ménage numbers' U_n:

The number of those normalized $3 \times n$ Latin rectangles with entries $1, 2, \ldots, n$ which are of the form

$$\begin{pmatrix} 1 & 2 & 3 & \ldots & n \\ n & 1 & 2 & \ldots & n-1 \\ a_{31} & a_{32} & a_{33} & \ldots & a_{3n} \end{pmatrix} \qquad (1.52)$$

is equal to

$$U_n = n! - \frac{2n}{2n-1} \binom{2n-1}{1} (n-1)! + \frac{2n}{2n-2} \binom{2n-2}{2} (n-2)! - \cdots$$

$$+ (-1)^n \frac{2n}{n} \binom{n}{n} 0! \quad \text{for} \quad n \geqslant 3.$$

The numbers U_n are called *ménage numbers* because they are the solutions of the following 'Problème des ménages' stated by É. Lucas:

Find the number of ways of seating n married couples at a round table with men and women in alternate positions so that no wife sits next to her husband.

The above formula for U_n was established by Touchard in 1934. In 1943 Kaplansky gave an elegant proof of Touchard's formula. Kaplansky's proof will be reproduced in part 2 of this book (see p. 366); the next two problems will be used as steps in the process of finding U_n.

Problem 132

In Lucas' problem of the married couples consider a fixed seating arrangement for the women. Prove that in this case the number of admissible seating arrangements for the men is equal to the number of normalized $3 \times n$ Latin squares of type (1.52).

Problem 133

(a) Prove that the number of ways of selecting objects, no two consecutive, from n objects arranged in a row, is

$$\binom{n-k+1}{k}.$$

(b) Prove that the number of ways of selecting k objects, no two consecutive, from n objects arranged in a circle, is

$$\frac{n}{n-k}\binom{n-k}{n}.$$

The next problem describes a connection between Latin squares and finite groups.

Problem 134

Let $G = \{g_1, g_2, \ldots, g_n\}$ be a finite group with respect to an operation $\circ$. The operation table of G is an $n \times n$ square array in which the entry in the ith row and in the jth column is the group element $g_i \circ g_j$ for all $i, j = 1, 2, \ldots, n$.
Prove that the operation table of G is a Latin square of order n.

The last problem in this section extends the notion of a Latin square to an array with infinitely many rows and columns.

Problem 135 (Kvant M1123)

A right angle is divided into infinitely many squares of side length a, as shown in Fig. 1.55. Is it possible to place into each square a natural number so that every row of squares and every column of squares contains any natural number exactly once?

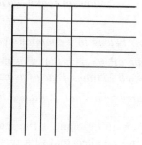

Fig. 1.55

10.2 Complete sets of mutually orthogonal Latin squares and finite affine and projective planes

Although Latin squares emerged as recreational curiosities, they have gained importance in applied and theoretical mathematics. Already in 1788 de Palluel used a 4×4 Latin square for the layout of a statistical experiment. Nowadays the construction of *complete set of mutually orthogonal Latin squares* is crucial in the study of finite geometries, such as *finite affine* and *projective planes*. The aim of this section is to point out the connection between these structures. First, a few definitions and problems.

A collection of distinct Latin squares $A_1, A_2, \ldots, A_k$ of order n is called a set of mutually orthogonal Latin squares, if A_i and A_j are orthogonal for all $i, j = 1, \ldots, k$, $i \neq j$.

It is not difficult to find an upper bound for the number of Latin squares of order n in a collection of mutually orthogonal Latin squares.

Problem 136

Prove that if $A_1, A_2, \ldots, A_k$ is a set of k orthogonal Latin squares of order $n \geqslant 3$, then $k \leqslant n - 1$.

Thus $n - 1$ is the greatest possible number of mutually orthogonal Latin squares of order n. *A set of $n - 1$ mutually orthogonal Latin squares of order n is called a complete set of mutually orthogonal Latin squares of order n.*

An affine plane $\mathcal{A}$ is a set $\mathcal{P}$ of elements, called points with certain subsets of $\mathcal{P}$ called lines such that the following requirements (called the axioms of $\mathcal{A}$) hold:

($\mathbf{A}_1$) *any two distinct points belong to a common line*

($\mathbf{A}_2$) *for any point–line pair (P, ℓ) such that P is not contained in ℓ, there is exactly one line ℓ' which contains P and has no common point with ℓ;*

($\mathbf{A}_3$) *there exist at least three points not contained in a common line.*

A classical example of an affine plane is the real euclidean plane Σ_2. In Chapter VIII, Section 8.1 we recalled that in a two-dimensional coordinate system in Σ_2, its points are represented by ordered pairs (x, y) or real numbers and its lines by the point sets (x, y) satisfying equations of the form

$y = mx + n$, where m and n are arbitrary real numbers,

or

$x = a$, where a is an arbitrary real number.

The above description of Σ_2 suggests a method for constructing other examples of affine planes:

Problem 137

(a) Consider the set $\mathbb{Z}_5 = \{0, 1, 2, 3, 4, 5\}$ of the remainders of integers when divided by 5. On $\mathbb{Z}_5$ define two binary operations: addition modulo 5 (denoted by $\oplus$) and multiplication modulo 5 (denoted by $\otimes$):

$a \oplus b$ is the remainder of the sum $a + b$ of a, $b \in \mathbb{Z}_5$ when divided by 5

$a \otimes b$ is the remainder of the product ab of a, $b \in \mathbb{Z}_5$ when divided by 5.

Define $\mathcal{A}_5$ as a structure of 'points' and 'lines' such that the points of $\mathcal{A}_5$ are the ordered pairs (x, y) where $x, y \in \mathbb{Z}_5$, and the lines of $\mathcal{A}_5$ are the sets of points (x, y) which satisfy an equation of the form

$$y = m \otimes x \oplus n \quad \text{for} \quad m, n \in \mathbb{Z}_5$$

or

$$x = a \qquad \text{for} \quad a \in \mathbb{Z}_5.$$

Prove that $\mathcal{A}_5$ is an affine plane.

(b) Prove that for any prime number p, the construction of $\mathcal{A}_5$, based on $\mathbb{Z}_5$ ($\oplus$, $\otimes$), can be generalized to the construction of an affine plane $\mathcal{A}_p$, based on $\mathbb{Z}_p = \{0, 1, 2, \ldots, p - 1\}$.

(c) Construct $\mathcal{A}_6$, based on $\mathbb{Z}_6$ ($\oplus$, $\otimes$), following the method in (a). Is $\mathcal{A}_6$ an affine plane?

The affine planes $\mathcal{A}_p$ have finitely many points, while Σ_2 has infinitely many points.

Now consider any affine plane A, that is, any structure satisfying axioms of (A_1)–(A_3). We impose on A the additional condition that one of its lines contain finitely many, say n, points. Then the following holds.

Problem 138

Let A be an affine plane.

If a line ℓ of A consists of n points, prove that:

(a) All lines of A have n points.

(b) All points of A belong to $n + 1$ lines.

(c) A has n^2 points.

(d) A has $n^2 + n$ lines.

(e) Any line of A belongs to a unique set of n lines such that no two lines of this set have a point in common.

The statements in Problem 138 lead to the following definitions.

An affine plane with finitely many points is called a *finite affine* plane.

The number of points on a line of a finite affine plane is the *order* of the affine plane.

Two lines ℓ and ℓ' of an affine plane are called *parallel* to one another if either $\ell = \ell'$, or ℓ and ℓ' have no point in common.

The set of all lines in an affine plane parallel to a given line of the plane forms a *parallel class* of the plane.

According to (d) and (e) in Problem 138, an affine plane of order n has $n + 1$ parallel classes.

Finite affine planes are linked with complete sets of orthogonal Latin squares through the following phenomenon.

Theorem T: A complete set of mutually orthogonal Latin squares of order n exists if and only if there exists an affine plane of order n.

The proof of Theorem T consists of two parts.

Part 1. Suppose there exists an affine plane $\mathcal{A}$ of order n.

Denote the $n + 1$ parallel classes by $\overline{R}, \overline{1}, \overline{2}, \ldots, \overline{n-1}$ and $\overline{C}$. In each parallel class label the lines $1, 2, \ldots, n$. Applying the axioms of an affine plane we can construct a square array A_k, corresponding to the parallel class $\overline{k}$, for $k = 1, 2, \ldots, n - 1$, as follows.

Problem 139

(a) Prove that the ith line (that is, the line labelled i) of $\overline{R}$ meets the jth line of $\overline{C}$ in a unique point of $\mathcal{A}$ for all $i, j = 1, 2, \ldots, n$. Denote this point by (i, j).

(b) Consider now an arbitrary parallel class $\overline{k}$. Verify that the point (i, j), defined in (a), belongs to exactly one of the lines from $\overline{k}$ (Fig. 1.56 (a)). Construct an $n \times n$ array A_k, corresponding to the parallel class $\overline{k}$, so that for all $i, j \in \{1, 2, \ldots, n\}$ the entry of A_k in the ith row and in the jth column is the label of that line of the parallel class $\overline{k}$ which contains the point (i, j). (In Fig. 1.56 (a) this label is t) (see Fig. 1.56 (b)).

In this way $n - 1$ square arrays $A_1, A_2, \ldots, A_{n-1}$ are constructed, corresponding to the parallel classes $\overline{1}, \overline{2}, \ldots, \overline{n-1}$.

(c) From the properties of the affine plane A deduce that:

(i) the square array A_k is a Latin square of order n for $k = 1, 2, \ldots, n - 1$; moreover

(ii) $A_1, A_2, \ldots, A_{n-1}$ form a complete set of mutually orthogonal Latin squares of order n.

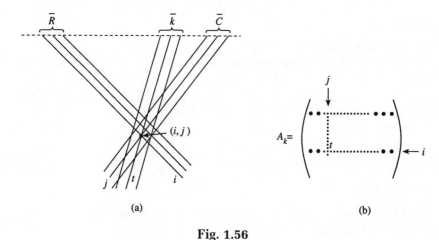

Fig. 1.56

This settles Part 1 of the proof of theorem T. We have shown that the existence of an affine plane implies the existence of a complete set of mutually orthogonal Latin squares of order n.

Part 2 of the proof of Theorem T is the subject of the next problem.

Problem 140

Suppose that there exists a complete set of mutually orthogonal Latin squares of order n. By reversing the method described in Problem 139, prove that this implies the existence of an affine plane of order n.
   Thus Theorem T is proved.

   Now we introduce the notion of a *projective plane*.

   A projective plane π is a set $\mathscr{P}^*$ of points with certain subsets of $\mathscr{P}^*$, called lines, satisfying the following axioms:

($\mathbb{P}_1$) *any two distinct points belong to a unique line;*

($\mathbb{P}_2$) *any two distinct lines have a unique point in common;*

($\mathbb{P}_3$) *there exist four points of $\mathscr{P}^*$ such that no three of them belong to a common line.*

   The connection between projective and affine planes is explained by the statements in Problems 141 and 142.

Problem 141

Let $\mathcal{A}$ be an affine plane.
Define $\mathcal{A}^*$ as a set of points and lines as follows:
The points of $\mathcal{A}^*$ are the points of $\mathcal{A}$ and the parallel classes of $\mathcal{A}$.
The lines of $\mathcal{A}^*$ are the lines of $\mathcal{A}$ and an additional line, consisting of those points of $\mathcal{A}^*$ which are the parallel classes of $\mathcal{A}$.
Prove that $\mathcal{A}^*$ is a projective plane.

Conversely:

Problem 142

Let π be a projective plane, and let ℓ be an arbitrary line of π.
Consider the structure π_ℓ, obtained from π as follows:
The points of π_ℓ are those points of π which do not belong to ℓ.
The lines of π_ℓ are the point sets obtained from the lines of π different from ℓ by removal of their points on ℓ.
Prove that π_ℓ is an affine plane.

From the assertions in Problems 141 and 142 it follows that the existence of affine planes of order n is equivalent to the existence of projective planes with $n + 1$ points on their lines. Such projective planes are called projective planes of order n. In view of theorem T, this implies that:

(T′) *A projective plane of order n exists if and only if there exists a complete set of mutually orthogonal Latin squares of order n.*

For which values of the natural number n does a projective plane of order n exist?

This is a very difficult question. Projective planes of order n have been constructed for all $n = p^\alpha$ where p is an arbitrary prime and α is an arbitrary natural number. No projective plane of order $n \neq p^\alpha$ has been found.

10.3   Problems on map colouring

In 1852 Francis Guthrie, a graduate student, noticed that on the map of England the counties can be coloured with four colours in such a way that neighbouring counties have different colours. Through his brother he asked Augustus de Morgan (1806–71) whether or not four colours would suffice for all maps. De Morgan thought that four colours would be sufficient but could not prove it. In 1878, at a meeting of the London Mathematical Society, Arthur Cayley (1821–95) asked if anyone could solve the problem. These were the origins of the four colour conjecture.

The beauty of the four colour conjecture is that it can be formulated in various equivalent ways, and therefore attempts to verify it promoted research in different fields of mathematics, for example in graph theory.

It was not until 1976 that the four colour problem was solved. Kenneth Appel and Wolfgang Haken from the University of Illinois employed computers for over 1000 hours of computer time to sort out various difficult cases and managed to prove that

Four colours are sufficient to colour any geographical map so that neighbouring countries have different colours.

The study of the four colour problem produced a number of substantially simpler questions and exercises on colouring special types of maps. We shall present some relevant examples, based on Dynkin and Uspenskii [37] and Golovina and Yaglom [38]. First some definitions.

A *map*, its *vertices, boundaries,* and *countries* are defined as follows.

A *map* is a network (in mathematical terms a graph) of line segments in a plane which connect certain points $A_1, A_2, \ldots, A_v$ and have no other points in common. Moreover, the network has to be *connected*, that is: for any $i, j \in \{1, 2, \ldots, v\}$ one can start at A_i and reach A_j by moving only along the lines of the network.

The points $A_1, \ldots, A_v$ are the *vertices* of the map, the line segments $A_i A_j$ joining the vertices A_i and A_j are the *boundaries*, and the regions of the plane into which it is divided by the boundaries are the *countries*. (The infinite region outside the boundaries will also be considered a country.)

An example of a map with countries $c_1, c_2, \ldots, c_6$ is shown in Fig. 1.57.

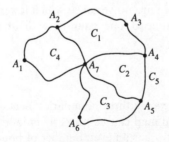

Fig. 1.57

We shall say that a map is coloured *properly* if each of its countries has a definite colour and any two countries with a common boundary have different colours.

Problem 143

Given n straight lines in a plane, prove that no matter how these lines are arranged, the map which they form can be properly coloured with two colours.

Problem 144

Given n circles in a plane, prove that no matter how they are arranged, the map which they form can be properly coloured with two colours.

The next statement can be called the 'two-colour theorem'.

Problem 145

Prove that a map can be properly coloured with *two* colours if and only if the number of the boundaries meeting at any vertex of the map is even.

Various questions on map colouring can be solved with the help of the following result obtained by Euler.

Problem 146 (Euler's theorem)

In an arbitrary map (having at least one vertex) let v denote the number of vertices, e the number of boundaries, and r the number of countries.
Prove that v, e, and r satisfy the formula

$$v - e + r = 2. \tag{1.53}$$

Problem 147

Prove that if a map contains no vertex at which fewer than three boundaries meet, then there is at least one country with at most five boundaries. (*Hint.* Use (1.53).)

Problem 148

Call a map *normal* if exactly three boundaries meet at each vertex.
Prove that a normal map can be properly coloured with *three* colours if and only if each country has an even number of boundaries.

In 1880 P. J. Heawood proved the five-colour theorem, stating that *five* colours are sufficient for colouring any map so that no neighbouring countries have the same colour. For a proof of this statement and for other interesting problems on map colouring see Dynking and Uspenskii [36].

We end this section with a remark on 'map colouring' in higher-dimensional spaces. At the turn of this century M. Crum proposed the following problem:

What is the maximum number of non-overlapping convex polyhedra in three-dimensional space such that any pair of them have a common boundary of positive area?

Surprisingly, the answer to the above question states that there is no such maximum number. In 1947, A. S. Besicovitch constructed an infinite sequence of convex polyhedra satisfying the required conditions. (Besicovitch was unaware of an earlier (1905) solution of Crum's problem by H. Tietze.) Besicovitch's result was generalized to n-dimensional spaces for $n \geqslant 3$ by R. Rado.

Thus in spaces of dimension $n > 2$ there are maps which cannot be coloured by any finite number of colours so that countries, meeting along boundaries of positive area, have different colours.

10.4   Problems related to graphs; memory wheels, and instant insanity

Graphs are used extensively in various disciplines, such as mathematics, communication theory, the natural and social sciences, and engineering, as well as in the solutions of many puzzles and exercises.

In this section we shall employ graphs to construct 'memory wheels' (number patterns encountered in information theory) and to solve the puzzle with the curious name 'instant insanity'.

We start by recalling a few definitions.

A *graph* is a set of points, called *vertices*, and a set of *edges*. Each edge joins two vertices, called the *endpoints* of the edge. There may be no edge joining certain two vertices, or one edge, or more. An edge is called a *loop* if its endpoints coincide (Fig. 1.58(a)).

A *path* of a graph is a succession of edges $a_1, a_2 \ldots, a_{k+1}, \ldots$ connecting a succession of vertices $P_1, P_2, \ldots, P_k, \ldots$, that is, a_1 connects P_1 and P_2, a_2 connects P_2 and $P_3, \ldots, a_k$ connects P_k and $P_{k+1}, \ldots$. We say that the path passes through $P_1, P_2 \ldots$.

Two vertices of a graph are said to be *connected* if there is a path passing through them (Fig. 1.58(b)).

A graph is called *connected* if every pair of its vertices is connected (Fig. 1.58(c)).

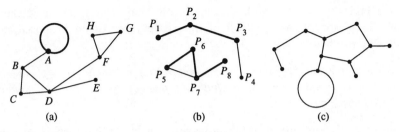

Fig. 1.58

An edge of a graph is said to be *oriented* if there is a distinction between its endpoints: one of them is the *beginning*, the other the *end* of the edge.

A graph is called *oriented* if all its edges are oriented. In diagrams of oriented graphs the orientation of the edges is shown by arrows on the edges, pointing from the beginning to the end of the edges (Fig. 1.59).

An *oriented circuit* is a path of an oriented graph with successive oriented edges $P_1 P_2, P_2 P_3, \ldots, P_{k-1} P_k$ such that P_k coincides with P_1 (e.g. $P_1 P_2, P_2 P_3, P_3 P_4, P_4 P_4, P_4 P_1$ form an oriented circuit of the graph in Fig. 1.59).

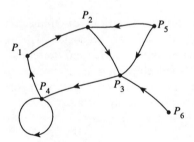

Fig. 1.59

An *Eulerian circuit* is a circuit of a connected oriented graph which contains each edge of the graph exactly once.

Eulerian circuits play an important role in the construction of so-called memory wheels

A *memory wheel* is a set of successive numbers $a_1, a_2, a_3, \ldots, a_n$ arranged in a circle (Fig. 1.60(a)). Usually each of the numbers a_i is either 0 or 1. Thus:

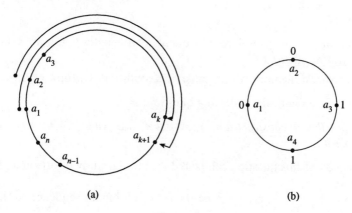

(a)   (b)

Fig. 1.60

For a fixed natural number $k \leqslant n$ each of the sequences

$$s_1 = a_1, a_2, \ldots, a_k$$
$$s_2 = a_2, a_3, \ldots, a_{k+1}$$
$$s_3 = a_3, a_4, \ldots, a_{k+2}$$
$$\cdot \qquad \cdot \qquad \cdot$$
$$s_{n-1} = a_{n-1}, a_n, \ldots, a_{k-2}$$
$$s_n = a_n, a_1, \ldots, a_{k-1}$$

is a binary sequence (that is, a sequence of zeros and ones) of length k.

The purpose of the memory wheel is to assist in recalling, as quickly as possible, any binary sequence of length k.

The above task can be achieved under the following condition.

(C) *Every binary sequence of length k is equal to exactly one of the sequences $s_1, s_2, \ldots, s_n$.*

Figure 1.60 (b) shows a memory wheel for $k = 2$. It satisfies condition (C) because it lists each of the four binary sequences of length 2: 0,0; 0,1; 1,0; 1,1:

$$a_1 a_2 \qquad\qquad = 0\,0$$
$$a_2 a_3 \qquad = \quad 0\,1$$
$$a_3 a_4 \quad = \qquad 1\,1$$
$$a_4 a_1 = \qquad\qquad 1\,0$$

Problem 149

(a) Find the number of binary sequences of length k for an arbitrary natural number k.

(b) Construct a memory wheel, satisfying condition (C) for $k = 3$.

We are faced with the following question:

Is it possible to construct a memory wheel, satisfying (C) for any natural number k?

The study of this question can be reduced to the study of oriented graphs as follows.

Let $S = \{s'_1, s'_2, \ldots, s'_i, \ldots, s'_j\}$ be the set of all binary sequences of length $k - 1$.

Construct an oriented graph G_k such that

the vertices of G_k are labelled $s'_1, s'_2, \ldots, s'_i, \ldots, s'_j$ (thus, each vertex of G corresponds to exactly one element of S);

two vertices s'_i and s'_j are connected by an oriented edge with beginning s'_i and end s'_j if and only if the last $k - 2$ terms of the sequence s'_i are the first $k - 2$ terms of the sequence s'_j (listed in the same order) — that is, if and only if

$$s'_i = a_{i_1} a_{i_2} \ldots a_{i_{k-1}} \quad \text{and} \quad s'_j = a_{i_2} a_{i_3} \ldots a_{i_k}$$

This oriented edge will be denoted by the binary sequence of length k:

$$s_{i_1} = a_{i_1} a_{i_2} \ldots a_{i_{k-1}} a_{i_k}.$$

Problem 150

Prove that a memory wheel satisfying (C) for an arbitrary natural number k can be constructed if and only if the graph G_k has an Eulerian circuit.

According to a result obtained by Euler it is known that

(E) *A connected, oriented graph has an Eulerian circuit if and only if for any vertex A of the graph the number of edges with beginning A is equal to the number of edges with end A.*

For a proof of (E) see Remark 1 of this section.

Problem 151

(a) Applying (E) and Problem 150 prove that it is possible to construct a memory wheel satisfying (C) for all $k = 1, 2, 3, \ldots$

(b) Carry out a construction for a memory wheel for $k = 4$.

Problem 152

Is it possible to construct a memory wheel containing

(a) all sequences of length k with digits $0, 1, 2, \ldots, 9$,
(b) all sequences of length $k = 3$ with *distinct* digits, belonging to $\{1, 2, \ldots, n\}$?

<u>*Remark*</u> *1.* Euler's theorem (E) can be proved by induction. What follows is a persuasive argument rather than a formal proof.

(a) Suppose that G is a connected, oriented graph such that for any vertex A of G the number of edges with beginning A is the same as the number of edges with end A.

   Then we can construct an Eulerian circuit of G in a number of steps.

Step 1. We start at any vertex A_1 and traverse edges following their orientation (once only) until we get stuck. This will happen when we return to A_1 after having traversed all edges through A_1. At this stage the traversed edges form a circuit c_1 which does not necessarily contain all edges of G (Fig. 1.61).

Step 2. Suppose that G has an edge e which is not in c_1. Since the graph G is connected, there exists a path p connecting e to c_1. Let A_2 be a vertex at which p meets c_1. Starting at A_2 we traverse edges of G, not included in c_1, until we get stuck at A_2. The edges traversed in step 2 form a circuit c_2 (Fig. 1.61).

Step 3. The circuits c_1 and c_2 can be combined into a single circuit c' which starts at A_2, traverses c_1, and then traverses c_2.

Step 4. If there is an edge not contained in c', then we repeat the same procedure as in step 2 and extend c' into a larger circuit c'', as in step 3.

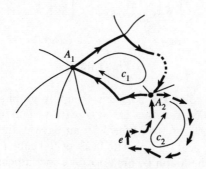

Fig. 1.61

The procedure in step 4 is repeated until all edges of G are included in a single circuit. This circuit is an Eulerian circuit of G.

(b) It is easily shown that if an oriented graph G has an eulerian circuit then it must be connected, and that, for any of its vertices, the number of edges ending at a vertex is equal to the number of edges starting at that vertex.

Remark 2. A circuit of a graph is called Hamiltonian (in honour of William Rowan Hamilton (1805–65)) if it passes through every *vertex* of the graph exactly once. Examples of graphs with or without Hamiltonian circuits are included in the next problem.

Problem 153

(a) Let S be a Platonic solid (that is, a regular tetrahedron, a cube, an octahedron, a dodecahedron, or an icosahedron). Let G be the graph whose vertices are vertices of S and whose edges are the edges of S. Find out whether G has a Hamiltonian circuit.

(b) Let S be a rhombic dodecahedron (see Chapter VII, Section 7.1), and let G be a graph whose vertices and edges are respectively the vertices and edges of S. Does G have a Hamiltonian circuit?

Instant insanity is a puzzle consisting of four cubes whose faces are coloured red, blue, green, and white, so that each cube has at least one face of each colour. The problem is to arrange these cubes on top of one another so that all four colours appear on each rectangular face of the resulting cuboid. Figure 1.62 shows the nets of four cubes forming the puzzle.

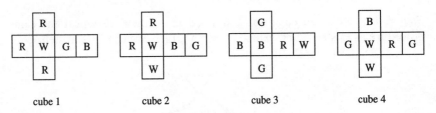

Fig. 1.62

A quick way to solve the puzzle is to apply graphs. Each cube i (for $i = 1, 2, 3, 4$) is associated with a graph G_i with vertices labelled by the four colours; two vertices of G_i are joined by an edge if and only if the corresponding colours appear on opposite faces of the cube i (Fig. 1.63). Figure 1.64 shows the graph G, obtained by superimposing G_1, G_2, G_3, and G_4.

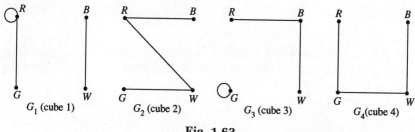

Fig. 1.63

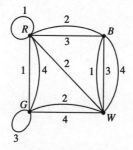

Fig. 1.64

Problem 154

Use Fig. 1.64 to solve the instant insanity puzzle.

Part 2

Solutions

Chapter I

Problem 1

Denote by p_n the number of different ways of painting a house with n floors, in accordance with the restrictions of the problem. p_n is the sum of the number of ways of painting that involve a white ground floor (w_n) and of the number of ways of painting that involve a blue ground floor (b_n):

$$p_n = w_n + b_n.$$

The colour of the floor above a white ground floor can be white or blue; therefore $w_n = p_{n-1}$. On the other hand, the colour of the floor above a blue ground floor must be white; hence $b_n = p_{n-2}$. Thus p_n satisfies the recurrence relation

$$p_n = p_{n-1} + p_{n-2} \quad \text{for} \quad n \geqslant 3.$$

Direct inspection shows that $p_1 = 2$ and $p_2 = 3$.

The above relations imply that the numbers $p_1, p_2, p_3, \ldots$ form part of the Fibonacci sequence: $p_n = f_{n+1}$ for $n = 1, 2, \ldots$.

Problem 2

Denote by s_n the number of different ways of painting a house with n floors. s_n is the sum of the number of ways of painting that involve a white ground floor (s_n'), of the number of ways of painting that involve a blue ground floor and a white floor above it (s_n''), and of the number of ways of painting that involve a blue ground floor and a blue floor above it (s_n'''):

$$s_n = s_n' + s_n'' + s_n'''.$$

The restrictions of the problem imply that $s_n' = s_{n-1}$, $s_n'' = s_{n-2}$, while $s_n''' = s_{n-3}$ (since a floor above two blue floors must be white). Thus s_n satisfies the recurrence relation

$$s_n = s_{n-1} + s_{n-2} + s_{n-3} \quad \text{for} \quad n \geqslant 4.$$

It is easy to verify that $s_1 = 2$, $s_2 = 4$ and $s_3 = 7$.

Thus the numbers s_n form the sequence $2, 4, 7, 13, 24, \ldots$ which is part of the 3-bonacci sequence $1, 1, 2, 4, 7, 13, 24, \ldots$ (see Definition (I), p. 4).

125

Problem 3

Denote by r_n the number of different ways of painting a house with n floors. r_n is the sum of the number of ways of painting that involve a white ground floor (r'_n), a grey ground floor (r''_n), and a blue ground floor (r'''_n). Since the colour of a floor above a white or a grey floor can be any of the three colours (white, grey, or blue) it follows that $r'_n = r''_n = r_{r-1}$. On the other hand, the colour above a blue ground floor must be white or grey; thus $r'''_n = 2r_{n-2}$. Hence

$$r_n = 2r_{n-1} + 2r_{n-2} \quad \text{for} \quad n \geqslant 3.$$

Direct inspection shows that $r_1 = 3$ and $r_2 = 8$.

Thus the numbers r_n form the sequence $3, 8, 22, 60, 164, \ldots$, which is the sequence $F(2, 2)$ with initial value $3, 4$, as in Definition (III) (p. 4).

Problem 4

Denote by $\hat{b}_n$ the total number of blue floors in the set of all differently painted n-floored houses. $\hat{b}_n$ is the sum of the number of blue floors in all houses with white ground floor ($\hat{b}'_n$), and of the number of blue floors in all houses with blue ground floor ($\hat{b}''_n$):

$$\hat{b}_n = \hat{b}'_n + \hat{b}''_n.$$

Clearly, $\hat{b}'_n = \hat{b}_{n-1}$. On the other hand, $\hat{b}''_n$ is the sum of the number of blue floors on the ground level ($\hat{b}'''_n$) and of the number of the remaining blue floors in houses with blue ground floors ($\hat{b}_{n-2}$). According to the solution of Problem 1, the number $\hat{b}'''_n$ is the Fibonacci number f_{n-1}. Thus $\hat{b}_n$ satisfies the recurrence relation

$$\hat{b}_n = \hat{b}_{n-1} + \hat{b}_{n-2} + f_{n-1} \quad \text{for} \quad n \geqslant 3.$$

Direct inspection shows that $\hat{b}_1 = 1$ and $\hat{b}_2 = 2$. (The numbers $\hat{b}_n$ form the 'Fibonacci sequence of order 1', defined in Section 1.4.)

Problem 5

All parts of this problem can be solved by using induction: Each of the statements (a)–(d) must be verified for the corresponding initial value of n; after that, under the assumption that the statement is true for an arbitrary value k of n, its truth must be proved for $n = k + 1$.

(a) $f_0 = f_2 - 1$, since $f_0 = 1$ and $f_2 = 2$.
　　Assume that $f_0 + f_1 + \ldots + f_{k-1} = f_{k+1} - 1$. In that case

$$f_0 + f_1 + \ldots + f_{k-1} + f_k = f_{k+1} - 1 + f_k.$$

However, $f_{k+1} + f_k = f_{k+2}$. Thus

$$f_0 + f_1 + \ldots + f_k = f_{k+2} - 1.$$

This proves (a).

(b) $f_0 = f_1$, since both numbers are equal to 1.
Assume that $f_0 + f_2 + f_4 + \ldots + f_{2k} = f_{2k+1}$. Then

$$f_0 + f_2 + \ldots + f_{2k} + f_{2k+2} = f_{2k+1} + f_{2k+2}.$$

$f_{2k+1} + f_{2k+2} = f_{2k+3}$; this proves (b)

(c) $f_1 = f_2 - 1$, because $f_1 = 1$ and $f_2 = 2$.
Assume that $f_1 + f_3 + \ldots + f_{2k-1} = f_{2k} - 1$. Then

$$f_1 + f_3 + \ldots + f_{2k-1} + f_{2k+1} = f_{2k} - 1 + f_{2k+1}.$$

Since $f_{2k} + f_{2k+1} = f_{2k+2}$, we have proved (c) for all $n \geqslant 1$.

(d) $f_1 = 1$ and $f_2 = 2$; hence f_1 and f_2 are coprime.
Assume that f_k and f_{k+1} are coprime.
Suppose that f_{k+1} and f_{k+2} are *not* coprime. In that case they have a common factor, say $d > 1$. Thus $f_{k+1} = df'_{k+1}$ and $f_{k+2} = df'_{k+2}$, where f'_{k+1} and f'_{k+2} are natural numbers. This implies that the relation $f_k = f_{k+2} - f_{k+1}$ can be rewritten in the form

$$f_k = df'_{k+2} - df'_{k+1} = d(f'_{k+2} - f'_{k+1}).$$

Thus d divides f_k as well as f_{k+1}, contrary to the assumption that f_k and f_{k+1} are coprime.

Problem 6

(a) The relation

$$f_{n+m} = f_{n-1}f_{m-1} + f_n f_m \tag{2.1}$$

will be proved by induction on n and m. This will be done in two steps.

Step 1. We shall prove that (2.1) holds for a fixed natural number m and for all natural numbers n, by using induction on n. For $n = 1$ (2.1) is transformed into

$$f_{1+m} = f_0 f_{m-1} + f_1 f_m = 1 \cdot f_{m-1} + 1 \cdot f_m = f_{m-1} + f_m.$$

This relation is true, since it represents the recurrence formula for the Fibonacci numbers.

Assume that (2.1) is true for $n \geqslant k$. In that case

$$f_k f_{m-1} + f_{k+1} f_m = (f_{k-1} + f_{k-2}) f_{m-1} + (f_k + f_{k-1}) f_m$$

$$= \underbrace{f_{k-1} f_{m-1} + f_k f_m}_{} + \underbrace{f_{k-2} f_{m-1} + f_{k-1} f_m}_{}$$

$$= \qquad f_{k+m} \qquad + \qquad f_{k-1+m}$$

$$= f_{k+1+m}.$$

This proves that (2.1) is true for a fixed m and for $n = k + 1$; in other words, (2.1) holds for a fixed m and for all n.

Step 2 consists in proving that (2.1) holds for any natural number n and any natural number m. To this end, take an arbitrary natural number n. The relation (2.1) holds for n and $m = 1$ since $f_{n+1} = f_{n-1} f_0 + f_n f_1$.

Assume that (2.1) holds for n and any $m \leqslant \ell$.
This implies that

$$f_{n-1} f_\ell + f_n f_{\ell+1} = f_{n-1}(f_{\ell-1} + f_{\ell-2}) + f_n(f_\ell + f_{\ell-1})$$

$$= \underbrace{f_{n-1} f_{\ell-1} + f_n f_\ell}_{} + \underbrace{f_{n-1} f_{\ell-2} + f_n f_{\ell-1}}_{}$$

$$= \qquad f_{n+\ell} \qquad + \qquad f_{n+\ell-1}$$

$$= f_{n+\ell+1}.$$

Hence (2.1) is true for any n and for $\ell + 1$, that is (2.1) is true for all natural numbers n and m.

(b) If n is divisible by m then $n = tm$ for some natural number t. We shall prove that f_{m-1} divides f_{km-1} for all natural numbers k. This will be done by induction on k, and by applying formula (2.1) step by step.

If $k = 2$ then $f_{2m-1} = f_{m+m-1}$. According to (2.1)

$$f_{m+m-1} = f_{m-1} f_{m-2} + f_m f_{m-1} = f_{m-1}(f_{m-2} + f_m).$$

Thus f_{m-1} divides f_{2m-1}.

Assume that f_{m-1} divides f_{km-1} for some k, i.e. that $f_{km-1} = f_{m-1}s$ for some natural number s.

Consider now $f_{(k+1)m-1}$. In view of (2.1) and of the induction hypothesis,

$$f_{(k+1)m-1} = f_{km-1+m} = f_{km-2} f_{m-1} + f_{km-1} f_m$$

$$= f_{km-2} f_{m-1} + (f_{m-1}s) f_m$$

$$= f_{m-1}(f_{km-2} + s f_m).$$

Hence f_{km-1} is divisible by f_{m-1} for all natural numbers k. In particular, f_{m-1} divides f_{tm-1}, that is f_{m-1} divides f_{n-1}.

Problem 7

(a) Proof by induction on n:

For $n = 2$

$$A^2 = \begin{pmatrix} 1 & 1 \\ 1 & 0 \end{pmatrix} \cdot \begin{pmatrix} 1 & 1 \\ 1 & 0 \end{pmatrix} = \begin{pmatrix} 2 & 1 \\ 1 & 1 \end{pmatrix} = \begin{pmatrix} f_2 & f_1 \\ f_1 & f_0 \end{pmatrix}.$$

Assume that

$$A^k = \begin{pmatrix} f_k & f_{k-1} \\ f_{k-1} & f_{k-2} \end{pmatrix},$$

then

$$A^{k+1} = A^k \cdot A = \begin{pmatrix} f_k & f_{k-1} \\ f_{k-1} & f_{k-2} \end{pmatrix}\begin{pmatrix} 1 & 1 \\ 1 & 0 \end{pmatrix} = \begin{pmatrix} f_k + f_{k-1} & f_k \\ f_{k-1} + f_{k-2} & f_{k-1} \end{pmatrix}\begin{pmatrix} f_{k+1} & f_k \\ f_k & f_{k-1} \end{pmatrix}$$

Thus

$$A^n = \begin{pmatrix} f_n & f_{n-1} \\ f_{n-1} & f_{n-2} \end{pmatrix} \text{ for all natural numbers } n \geqslant 2.$$

(b) $f_{n+1} f_{n-1} - f_n^2$ is the determinant of the matrix A^{n+1}. It is known that for any two $k \times k$ matrices A and B (for $k = 1, 2, \dots$) the determinant of the product AB is the product of the determinants of A and B (verify this for $k = 2$). Hence

$$\text{Determinant } A^{n+1} = (\text{Determinant } A)^{n+1}.$$

It follows that

$$f_{n+1} f_{n-1} - f_n^2 = (-1)^{n+1} \quad \text{for} \quad n \geqslant 2. \tag{2.2}$$

Formula (2.2) is also true for $n = 1$ because

$$f_2 f_0 - f_1^2 = 2 \cdot 1 - 1^2 = 1 = (-1)^2.$$

Problem 8

(a) We use induction on n.

For $n = 0$

$$\frac{1}{\sqrt{5}}\left[\left(\frac{1 + \sqrt{5}}{2}\right)^1 - \left(\frac{1 - \sqrt{5}}{2}\right)^1\right] = \frac{1}{\sqrt{5}} \cdot \sqrt{5} = 1 = f_0.$$

Assume that

$$f_n = \frac{1}{\sqrt{5}} \left[\left(\frac{1 + \sqrt{5}}{2} \right)^{n+1} - \left(\frac{1 - \sqrt{5}}{2} \right)^{n+1} \right] \quad \text{for all } n \leqslant k.$$

Then, since $f_{k+1} = f_k + f_{k-1}$, it follows that

$$f_{k+1} = \frac{1}{\sqrt{5}} \left[\left(\frac{1 + \sqrt{5}}{2} \right)^{k+1} - \left(\frac{1 - \sqrt{5}}{2} \right)^{k+1} \right] + \frac{1}{\sqrt{5}} \left[\left(\frac{1 + \sqrt{5}}{2} \right)^{k} - \left(\frac{1 - \sqrt{5}}{2} \right)^{k} \right]$$

$$= \frac{1}{\sqrt{5}} \left[\left(\frac{1 + \sqrt{5}}{2} \right)^{k} \left(\frac{1 + \sqrt{5}}{2} + 1 \right) - \left(\frac{1 - \sqrt{5}}{2} \right)^{k} - \left(\frac{1 - \sqrt{5}}{2} + 1 \right) \right]$$

$$= \frac{1}{\sqrt{5}} \left[\left(\frac{1 + \sqrt{5}}{2} \right)^{k} \left(\frac{1 + \sqrt{5}}{2} \right)^{2} - \left(\frac{1 - \sqrt{5}}{2} \right)^{k} \left(\frac{1 - \sqrt{5}}{2} \right)^{2} \right]$$

$$= \frac{1}{\sqrt{5}} \left[\left(\frac{1 + \sqrt{5}}{2} \right)^{k+2} - \left(\frac{1 - \sqrt{5}}{2} \right)^{k+2} \right].$$

This proves Binet's formula for all integers $n \geqslant 0$.

(b) According to (a),

$$\frac{f_n}{f_{n-1}} = \frac{\frac{1}{\sqrt{5}} \left[((1 + \sqrt{5})/2)^{n+1} - ((1 - \sqrt{5})/2)^{n+1} \right]}{\frac{1}{\sqrt{5}} \left[((1 + \sqrt{5})/2)^{n} - ((1 - \sqrt{5})/2)^{n} \right]}$$

$$= \frac{1}{2} \frac{1 + \sqrt{5} - (1 - \sqrt{5})((1 - \sqrt{5}/(1 + \sqrt{5}))^{n}}{1 - ((1 - \sqrt{5})/(1 + \sqrt{5}))^{n}} \tag{2.3}$$

Since $|(1 - \sqrt{5})/(1 + \sqrt{5})| < 1$, the fraction $((1 - \sqrt{5})/(1 + \sqrt{5}))^{n}$ tends to 0 as n tends to infinity. Hence, in view of (2.3),

$$\lim_{n \to \infty} \frac{f_n}{f_{n-1}} = \frac{1 + \sqrt{5}}{2}.$$

Problem 9

(a) Suppose that there exist two natural numbers n and m, with $n > m$, such that the Fibonacci rectangles R_n and R_m are similar. This implies that their corresponding sides are proportional:

$$f_n : f_{n-1} = f_m : f_{m-1}.$$

Hence

$$f_n f_{m-1} = f_m f_{n-1}. \tag{2.4}$$

Using the recurrence formula for Fibonacci numbers we can rewrite (2.4) as

$$(f_{n-1} + f_{n-2}) f_{m-1} = (f_{m-1} + f_{m-2}) f_{n-1}. \tag{2.5}$$

This yields

$$f_{n-1} f_{m-2} = f_{n-2} f_{m-1}.$$

Similarly, (2.5) can be transformed into

$$(f_{n-2} + f_{n-3}) f_{m-2} = f_{n-2}(f_{m-2} + f_{m-3}).$$

This implies that

$$f_{n-2} f_{m-3} = f_{m-2} f_{n-3}.$$

In this way, step by step, one obtains the relations

$$f_{n-k} f_{m-(k+1)} = f_{m-k} f_{n-(k+1)} \text{ for all } k = 1, 2, \ldots, m - 1.$$

The last of these relations, for $k = m - 1$, is of the form

$$f_{n-m+1} f_0 = f_1 f_{n-m}. \tag{2.6}$$

Since $f_0 = f_1 = 1$, this would imply that $f_{n-m} = f_{n-m+1}$. The latter equality cannot be true, because the Fibonacci sequence contains only one pair of equal consecutive terms, namely f_0 and f_1. However, in view of our initial assumption $n > m$.

Thus (2.6) is false, which implies that (2.4) does not hold for any n, m where $n > m$. There are no similar Fibonacci rectangles R_n and R_m for $n \neq m$.

(b) According to Problem 8(b)

$$\lim_{n \to \infty} \frac{f_n}{f_{n-1}} = \frac{1 + \sqrt{5}}{2} = \frac{(1 + \sqrt{5})/2}{1}.$$

Thus, for large values of n, the ratio $f_n : f_{n-1}$ of the sides of the Fibonacci rectangle R_n approaches the ratio of the sides of a rectangle R_G, whose dimensions are $(1 + \sqrt{5})/2$ and 1.

Problem 10

(a) Figure 2.1 shows the Fibonacci rectangle $R_n = ABCD$ with base AB of length f_n and height BC of length f_{n-1} such that $n \geqslant 2$.

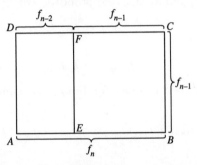

Fig. 2.1

Since the Fibonacci number f_n for $n \geqslant 2$ is the sum of the two preceding terms of the Fibonacci sequence, there is a unique point E on the base AB of R_n such that AE and EB have lengths f_{n-2} and f_{n-1} respectively. Through E draw a straight line, parallel to BC and intersecting CD at a point F. The straight line segment EF partitions R_n into the square $S_{n-1} = EBCF$ of side length f_{n-1} and the Fibonacci rectangle $R_{n-1} = AEFD$.

(b) From (a) it follows that

$$\text{Area } ABCD = \text{Area } AEFD + \text{Area } EBCF,$$

that is

$$f_n f_{n-1} = f_{n-1} f_{n-2} + f_{n-1}^2.$$

(c) The procedure of partitioning R_n into S_{n-1} and R_{n-1} can be iterated. The rectangle $AEFD$ can be partitioned into a square $S_{n-2} = GFDH$ of side length f_{n-2} and the Fibonacci rectangle $R_{n-2} = AEGH$ (Fig. 2.2).

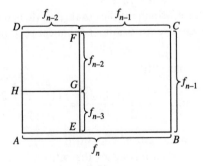

Fig. 2.2

By repeating this procedure, R_n is partitioned into $S_{n-1}, S_{n-2}, \ldots,$ S_1, R_1. The rectangle R_1 with side lengths $f_1 = 1$ and $f_0 = 1$ and the square S_0 of side length f_0 are identical.

(d) Follows in a straightforward manner from the above partition of R_n into squares which implies that

$$f_n f_{n-1} = \text{Area } R_n = \text{Area } S_{n-1} + \text{Area } S_{n-2} + \ldots + \text{Area } S_0$$

$$= f_{n-1}^2 + f_{n-2}^2 + \ldots + f_0^2.$$

Problem 11

(a) Figure 2.3 represents the 3-bonacci hexagon $P_n(3) = ABCDEF$ with side lengths $AB = f_n(3)$, $BC = f_{n-1}(3)$, and $CD = f_{n-2}(3)$, and angles of 120°. Since $f_n(3) = f_{n-1}(3) + f_{n-2}(3) + f_{n-3}(3)$, there is a unique point G on AB such that $BG = f_{n-3}(3)$ and $AG = f_{n-1}(3) + f_{n-2}(3)$. Similarly, there is a unique point H on ED such that $DH = f_{n-3}(3)$ and $HE = f_{n-1}(3) + f_{n-2}(3)$. Draw the straight line segment FI through F, parallel to AG (and to EH), such that $FI = AG = EH$, and join I to G and to H.

In this way $P_n(3)$ is partitioned into a hexagon $BCDHIG$ and a hexagon $P'_{n-1}(3) = AFEHIG$.

$BCDHIG$ is a 3-bonacci hexagon $P_{n-1}(3)$, since its angles are all 120° and its consecutive sides have lengths $f_{n-1}(3)$, $f_{n-2}(3)$, $f_{n-3}(3)$, $f_{n-1}(3)$, $f_{n-2}(3)$, and $f_{n-3}(3)$. The hexagon $P'_{n-1}(3)$ is concave; it consists of two parallelograms $AFIG$ and $FEHI$, with angles of 60° and 120° and with side lengths $f_{n-1}(3)$, $f_{n-1}(3) + f_{n-2}(3)$ and $f_{n-2}(3)$, $f_{n-1}(3) + f_{n-2}(3)$ respectively.

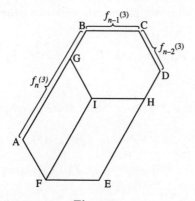

Fig. 2.3

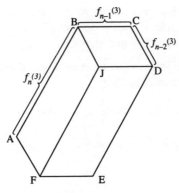

Fig. 2.4

(b) A 3-bonacci hexagon $P_n(3)$ can be partitioned into three parallelograms
   with angles of 60° and 120°, as shown in Fig. 2.4: $ABJF$ with sides $f_n(3)$
   and $f_{n-2}(3)$, and area $f_n(3) f_{n-2}(3) \sin 120°$; $FJDE$ with sides $f_n(3)$ and
   $f_{n-1}(3)$, and area $f_n(3) f_{n-1}(3)\sin 120°$; $BJDC$ with sides $f_{n-1}(3)$ and
   $f_{n-2}(3)$, and area $f_{n-1}(3) f_{n-2}(3) \sin 120°$.
   Thus the area of $P_n(3)$ can be written as

$$\text{Area } P_n(3) = \left[f_n(3) f_{n-1}(3) + f_n(3) f_{n-2}(3) + f_{n-1}(3) f_{n-2}(3) \right] \sin 120°.$$
$$(2.7)$$

According to this formula,

$$\text{Area } P_{n-1}(3) = \left[f_{n-1}(3) f_{n-2}(3) + f_{n-1}(3) f_{n-3}(3) + f_{n-2}(3) f_{n-3}(3) \right]$$
$$\sin 120°. \qquad (2.8)$$

The partition of $P_n(3)$ into $P_{n-1}(3)$ and $P'_{n-1}(3)$ (shown in part (a))
establishes the following relationship between the areas of $P_n(3)$ and
$P_{n-1}(3)$:

$$\text{Area } P_n(3) = \text{Area } P_{n-1}(3) + \text{Area } P'_{n-1}(3) \qquad (2.9)$$

where

$$\text{Area } P'_{n-1}(3) = \text{Area } AFIG + \text{Area } FEHI$$
$$= f_{n-1}(3)\left[f_{n-1}(3) + f_{n-2}(3) \right] \sin 120°$$
$$+ f_{n-2}(3)\left[f_{n-1}(3) + f_{n-2}(3) \right] \sin 120°$$
$$= \left[f_{n-1}(3) + f_{n-2}(3) \right]^2 \sin 120°. \qquad (2.10)$$

From relations (2.7)–(2.10) it follows that

$$f_n(3)f_{n-1}(3) + f_n(3)f_{n-2}(3) + f_{n-1}(3)f_{n-2}(3) = f_{n-1}(3)f_{n-2}(3)$$

$$+ f_{n-1}(3)f_{n-3}(3) + f_{n-2}(3)f_{n-3}(3) + \left[f_{n-1}(3) + f_{n-2}(3) \right]^2.$$

(2.11)

(c) The partitioning of $P_n(3)$ into $P_{n-1}(3)$ and $P'_{n-1}(3)$ can be iterated. Inside $P_n(3)$, following the method in (a),

partition $P_{n-1}(3)$ into $P_{n-2}(3)$ and $P'_{n-2}(3)$,

partition $P_{n-2}(3)$ into $P_{n-3}(3)$ and $P'_{n-3}(3)$,

....

partition $P_3(3)$ into $P_2(3)$ and $P'_2(3)$.

In this way the original 3-bonacci hexagon $P_n(3)$ is partitioned into $P'_{n-1}(3), P'_{n-2}(3), \ldots, P'_2(3)$, and $P_2(3)$. Hence the area of $P_n(3)$ is the sum of the areas of these polygons.

In view of (2.10), the area of $P'_i(3)$ is equal to $\left[f_i(3) + f_{i-1}(3) \right]^2 \sin 120°$ for all $i = 2, 3, \ldots, n - 1$. According to (2.7) the area of $P_2(3)$ is equal to $\left[f_2(3)f_1(3) + f_2(3)f_0(3) + f_1(3)f_0(3) \right] \sin 120°$, that is, to $\left\{ \left[f_1(3) + f_0(3) \right]^2 + \left[f_0(3) \right]^2 \right\} \sin 120°$. This leads to the formula

$$f_n(3)f_{n-1}(3) + f_n(3)f_{n-2}(3) + f_{n-1}(3)f_{n-2}(3)$$

$$= \sum_{i=1}^{n-1} \left[f_i(3) + f_{i-1}(3) \right]^2 + \left[f_0(3) \right]^2.$$

(2.12)

Problem 12

(i) In order to try to generalize (2.11) for $f_n(k)$, where $k \geqslant 2$, we consider the sum $S_n(k)$ of all products $f_i(k)f_j(k)$ for $i = n, n - 1, \ldots, n - (k - 2)$, and $j = n - 1, n - 2, \ldots, n - (k - 1)$, with $i > j$:

$$S_n(k) = f_n(k)f_{n-1}(k) + f_n(k)f_{n-2}(k) + f_n(k)f_{n-3}(k) + \ldots + f_n(k)f_{n-k+1}(k)$$

$$+ f_{n-1}(k)f_{n-2}(k) + f_{n-1}(k)f_{n-3}(k) + \ldots + f_{n-1}(k)f_{n-k+1}(k)$$

$$+ f_{n-2}(k)f_{n-3}(k) + \ldots + f_{n-2}(k)f_{n-k+1}(k)$$

....

$$+ f_{n-k+2}(k)f_{n-k+1}(k).$$

Our aim is to represent the difference $S_n(k) - S_{n-1}(k)$ for $n \geqslant 1$ as the square of a sum of k-bonacci numbers in a manner similar to (2.11). To this end recall that

$$f_n(k) = \sum_{i=1}^{k} f_{n-i}(k) \quad \text{for} \quad n \geqslant k.$$

Replacing $f_n(k)$ in $S_n(k)$ by the above sum we find that

$$S_n(k) = \sum_{i=1}^{k} f_{n-i}(k) \sum_{i=1}^{k-1} f_{n-i}(k) + f_{n-1}(k) \sum_{i=2}^{k-1} f_{n-i}(k)$$

$$+ f_{n-2}(k) \sum_{i=3}^{k-1} f_{n-i}(k) + \ldots + f_{n-k+2} \sum_{i=k-1}^{k-1} f_{n-i}(k)$$

By rearranging the right-hand side of this relation it follows that

$$S_n(k) = \left(\sum_{i=1}^{k-1} f_{n-i}(k) \right)^2 + f_{n-k}(k) \sum_{i=1}^{k-1} f_{n-i}(k) + f_{n-1}(k) \sum_{i=2}^{k-1} f_{n-i}(k)$$

$$+ f_{n-2}(k) \sum_{i=3}^{k-1} f_{n-i}(k) + \ldots + f_{n-k+2}(k) \sum_{i=k-1}^{k-1} f_{n-i}(k)$$

$$= \left(\sum_{i=1}^{k-1} f_{n-i}(k) \right)^2 + f_{n-1}(k) \sum_{i=2}^{k} f_{n-i}(k) + f_{n-2}(k) \sum_{i=3}^{k} f_{n-i}(k) + \ldots$$

$$+ f_{n-k+2}(k) \sum_{i=k-1}^{k} f_{n-i}(k) + f_{n-k+1}(k) \sum_{i=k}^{k} f_{n-i}(k).$$

In the above expression for $S_n(k)$ the summands following $\left(\sum_{i=1}^{k-1} f_{n-i}(k) \right)^2$ add up to $S_{n-1}(k)$.
Thus

$$S_n(k) - S_{n-1}(k) = \left[f_{n-1}(k) + f_{n-2}(k) + \ldots + f_{n-k+1}(k) \right]^2. \qquad (2.13)$$

(2.13) generalizes (2.11) for all k-bonacci numbers.

(ii) In (2.13) replace $S_{n-1}(k)$ by the expression $S_{n-2}(k) + \left[f_{n-2}(k) + f_{n-3}(k) + \ldots + f_{n-k}(k) \right]^2$; this gives

$$S_n(k) = S_{n-2}(k) + \left[f_{n-1}(k) + f_{n-2}(k) + \ldots + f_{n-k+1}(k) \right]^2$$

$$+ \left[f_{n-2}(k) + f_{n-3}(k) + \ldots + f_{n-k}(k) \right]^2 \qquad (2.14)$$

In (2.14) replace $S_{n-2}(k)$ by $S_{n-3}(k) + \left[f_{n-3}(k) + f_{n-4}(k) + \ldots + f_{n-k-1}(k) \right]^2$. Following this procedure we can express $S_n(k)$ as the following sum of squares:

$$S_n(k) = \sum_{i=0}^{n-1} \left[f_i(k) + f_{i-1}(k) + f_{i-2}(k) + \ldots + f_{i-k+2}(k) \right]^2, \qquad (2.15)$$

where $f_j(k) = 0$ for all negative integers j. Formula (2.15) generalizes (2.12) for all $k \geqslant 2$.

Problem 13

(a) Let O be the common point of the diagonals $A_0 A_2$ and $A_1 A_3$ (Fig. 2.5). Without loss of generality we can take $A_1 A_0 = 1$. In that case $A_1 A_2 = \varphi$ and $A_2 A_3 = \varphi^2$. The triangles $A_0 A_1 A_2$ and $A_1 A_2 A_3$ are right triangles

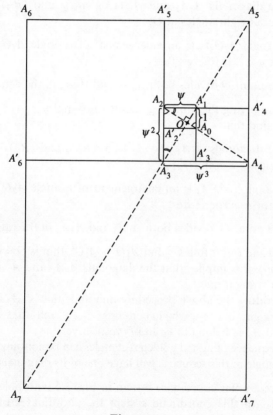

Fig. 2.5

and the ratio of their sides which form the right angle is $1:\varphi$. Hence triangle $A_1A_2A_3$ is an enlargement of triangle $A_0A_1A_2$ (the scale factor of this enlargement is φ). Therefore

$$\angle A_2A_3O = \angle A_1A_2A_0.$$

On the other hand,

$$\angle OA_2A_3 = \angle A_1A_0A_2 \text{ (alternate angles).}$$

The above equalities imply that

$$\angle OA_2A_3 + \angle A_2A_3O = \angle A_1A_0A_2 + \angle A_1A_2A_0 =$$
$$180° - \angle A_0A_1A_2 = 90°.$$

Therefore, from triangle A_2OA_3 it follows that

$$\angle A_2OA_3 = 180° - (\angle OA_2A_3 + \angle A_2A_3O) = 180° - 90° = 90°.$$

Thus

(i)   The diagonals A_0A_2 and A_1A_3 of $(R_G)_0$ and $(R_G)_1$ respectively meet at right angles at O.

Triangle A_2OA_3 is an enlargement of triangle A_1OA_0 with scale factor φ^2. It follows that

(ii)   The point O divides both A_0A_2 and A_1A_3 in the ratio $1:\varphi^2$.

By applying the same arguments to triangles $A_1A_2A_3$ and $A_2A_3A_4$ we find that

(iii)   The diagonals A_1A_3 and A_2A_4 of $(R_G)_1$ and $(R_G)_2$ meet at right angles at a point O'.

Triangle $A_3O'A_4$ is an enlargement of triangle $A_2O'A_1$ with scale factor φ^2. Therefore

(iv)   The point O' divides both A_1A_3 and A_2A_4 in the ratio $1:\varphi^2$.

From (ii) and (iv) it follows that $A_1O = A_1O'$, that is $O = O'$. In view of (i) and (iii), this implies that the diagonals A_0A_2 and A_2A_4 belong to a common straight line ℓ_1.

By iterating the above procedure one finds that A_3A_5 belongs to the same straight line ℓ_2, which contains A_1A_3, and that the diagonals A_4A_6 and A_5A_7 belong to ℓ_1 and ℓ_2 respectively.

The sequence of nested golden rectangles can be continued indefinitely. Each triangle in this sequence will have one of its diagonals on ℓ_1 or ℓ_2.

(b) Introduce a polar coordinate system in Fig. 2.5 with pole O, and polar axis OA_0. In this coordinate system the coordinates of an arbitrary point are denoted by r and ϑ.

From the similarity of the triangles $A_0OA_1, A_1OA_2, A_2OA_3, \ldots$ it follows that the polar coordinates of A_K for $k = 0, 1, \ldots, 6$ can be expressed in the form $r_k = r_0\varphi^k$ and $\vartheta_k = k\dfrac{\pi}{2}$ (in radians), where $r_0 = OA_0$.

Hence the coordinates (r_k, ϑ_k) satisfy the equation $r = r_0\varphi^{2\vartheta/\pi}$. Since

$$\frac{dr}{d\vartheta} = r\frac{2}{\pi} \ln \varphi,$$

i.e. $dr/d\vartheta$ is proportional to r, this is the equation of a logarithmic spiral σ (see p. 11).

Thus $A_0, A_1, \ldots, A_7$ belong to σ.

Problem 14

(a) Let O be the common point of the diagonals A_0A_3 and A_1A_4 of the silver hexagons $(P_S(3))_0$ and $(P_S(3))_1$ respectively (Fig. 2.6). Without loss of generality we can put $A_0A_1 = 1$. In that case $A_1A_2 = \varphi(3)$, $A_2A_3 = \varphi^2(3)$, and $A_3A_4 = \varphi^3(3)$.

$(P_S(3))_0$ and $(P_S(3))_1$ are similar, and therefore their corresponding parts, in particular the quadrilaterals $A_0A_1A_2A_3$ and $A_1A_2A_3A_4$, are also similar. Hence $\angle A_2A_3A_0 = \angle A_3A_4A_1$. This implies that

$$\angle OA_3A_4 + \angle A_3A_4O = \angle OA_3A_4 + \angle A_2A_3A_0 = 120°.$$

But then in triangle OA_3A_4, $\angle A_3OA_4 = 180° - 120° = 60°$. Thus

(i) A_0A_3 and A_1A_4 meet at O and form an angle of 60°. Triangle A_3OA_4 is an enlargement of triangle A_0OA_1 with scale factor $\varphi^3(3)$.

Therefore

(ii)   The point O divides both A_0A_3 and A_1A_4 in the ratio $1:\varphi^3(3)$.

By analogous arguments, applied to the silver hexagons $(P_S(3))_1$ and $(P_S(3))_2$, one deduces that

(iii) A_1A_4 and A_2A_5 meet at a point O' and form an angle of 60°, and that

(iv)   The point O' divides A_1A_4 and A_2A_5 in the ratio $1:\varphi^3(3)$.

(ii) and (iv) imply that $O' = O$.

Similarly, by considering $(P_S(3))_2$ and $(P_S(3))_3$, we can show that

(v) A_2A_5 and A_3A_6 meet at a point O'' and form an angle of 60°, and that

(vi)   The point O'' divides A_2A_5 and A_3A_6 in the ratio $1:\varphi^3(3)$.

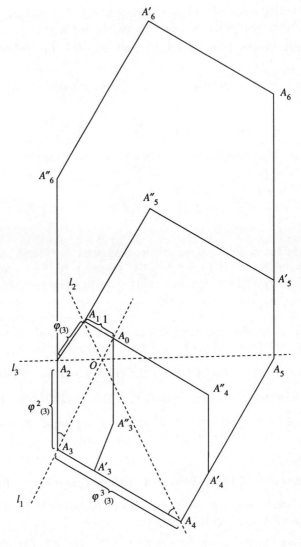

Fig. 2.6

From (iv) and (vi) it follows that $O'' = O$, while (i), (iii), and (v) imply that A_0A_3 and A_3A_6 belong to a common straight line ℓ_1.

The sequence of nested silver hexagons can be continued indefinitely. Using the method applied above one can prove that each of these hexagons has a diagonal which belongs to one of the straight lines $\ell_1 = A_0A_3$, $\ell_2 = A_1A_4$, and $\ell_3 = A_2A_5$ that meet at a common point O.

(b) Set up a polar coordinate system in Fig. 2.6 with pole O and polar axis OA_0. In this coordinate system the points A_k have coordinates $r_k = OA_k$ and $\vartheta_k = k\,\pi/3$ (in radians) for $k = 0, 1, 2, \ldots$ Since O divides $A_k A_{k+3}$ in the ratio $1{:}\varphi^3(3)$, it follows that $OA_k = (1/1 + \varphi^3(3))A_k A_{k+3}$. The similarity of the silver hexagons $(P_S(3))_k$ and $(P_S(3))_{k+1}$ with scale factor $\varphi(3)$, implies that $A_k A_{k+3} = A_0 A_3 [\varphi(3)]^k$. Hence

$$r_k = \frac{A_0 A_3}{1 + \varphi^3(3)}\,\varphi^k(3).$$

Thus r_k and ϑ_k satisfy the equation

$$r = \frac{A_0 A_3}{1 + \varphi^3(3)}\,[\varphi(3)]^{3\vartheta/\pi}.$$

This is the equation of a logarithmic spiral $\sigma(3)$. Hence $A_0, A_1, \ldots, A_6$ belong to $\sigma(3)$.

Problem 15

Figure 2.7 represents a 3-bonacci cuboid $C_n(3) = ABCDEFGH$ with edge lengths $AB = f_n(3)$, $BC = f_{n-1}(3)$, and $CG = f_{n-2}(3)$. The points B', C', G', and F' on the edges AB, DC, HG, and EF respectively are such that $AB' = DC' = HG' = EF' = f_{n-3}(3)$.

In this way $C_n(3)$ is partitioned into the 3-bonacci cuboid $C_{n-1}(3) = AB'C'DEF'G'H$ and the cuboid $C'_{n-1}(3) = B'BCC'F'FGG'$ with edge lengths $BC = f_{n-1}(3)$, $CG = f_{n-2}(3)$, and $B'B = f_n(3) - f_{n-3}(3) = f_{n-1}(3) + f_{n-2}(3)$. Hence

$$\text{Volume } C_n(3) = \text{Volume } C_{n-1}(3) + \text{Volume } C'_{n-1}(3),$$

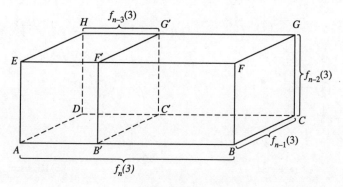

Fig. 2.7

that is

$$f_n(3)f_{n-1}(3)f_{n-2}(3) = f_{n-1}(3)f_{n-2}(3)f_{n-3}(3) + f_{n-1}(3)f_{n-2}(3)[f_{n-1}(3) + f_{n-2}(3)].$$

By iterating this procedure (that is, by partitioning $C_k(3)$ into $C_{k-1}(3)$ and $C'_{k-1}(3)$ for $k = n-1, n-2, \ldots, 3$) we obtain that

$$f_n(3)f_{n-1}(3)f_{n-2}(3) = \text{Volume } C_2(3) + \sum_{k=2}^{n-1} f_k(3)f_{k-1}(3)[f_k(3) + f_{k-1}(3)]$$

$$= 1 \cdot 1 \cdot 2 + \sum_{k=2}^{n-1} f_k(3)f_{k-1}(3)[f_k(3) + f_{k-1}(3)]$$

$$= \sum_{k=1}^{n-1} f_k(3)f_{k-1}(3)[f_k(3) + f_{k-1}(3)].$$

Problem 16

(a) Figure 2.8(a) recalls the method of construction of Pascal's triangle (described in Section 1.4). Each entry on the diagonal d_{n-1} is used twice as a summand for forming all entries of d_n. Thus for Pascal's triangle

$$\bar{d}_n = 2\bar{d}_{n-1} \quad \text{for} \quad n \geqslant 1; \quad \bar{d}_0 = 1.$$

(b) Figure 2.8(b) shows how to construct the Fibonacci triangle (explained in Section 1.4): Each entry on the diagonal d_{n-2} is used once, and each

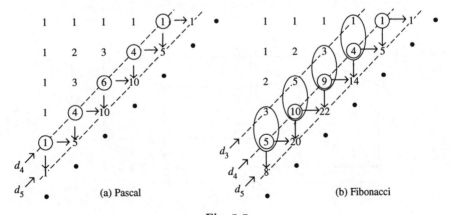

(a) Pascal   (b) Fibonacci

Fig. 2.8

entry on diagonal d_{n-1} is used twice to form all entries of d_n. Hence for the Fibonacci triangle

$$\bar{d}'_n = 2\bar{d}'_{n-1} + \bar{d}'_{n-2} \quad \text{for} \quad n \geq 2; \quad \bar{d}'_0 = 1 \quad \text{and} \quad \bar{d}'_1 = 2.$$

(c) Figure 2.9(a) shows how to construct the terms on the lines s_n: Each entry of s_{n-1} is used once and each entry of s_{n-2} is used once to form all entries of s_n. Therefore

$$\bar{s}_n = \bar{s}_{n-1} + \bar{s}_{n-2} \quad \text{for} \quad n \geq 2; \quad \bar{s}_0 = \bar{s}_1 = 1.$$

(In other words, the numbers $\bar{s}_n$ are the Fibonacci numbers f_n for $n = 0, 1, 2, \ldots$)

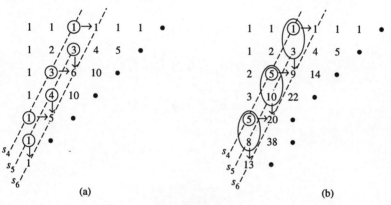

(a)   (b)

Fig. 2.9

(d) As shown in Fig. 2.9(b) each entry of s_{n-1} is used once, and each entry of s_{n-2} is used twice to form all entries of $\bar{s}_n$. Hence

$$\bar{s}'_n = \bar{s}'_{n-1} + \bar{s}'_{n-2} \quad \text{for} \quad n \geq 2; \quad \bar{s}'_0 = \bar{s}'_1 = 1.$$

Problem 17

(a) $$C_i^{k+i} = C_i^{(k-1)+i} + C_{i-1}^{k+(i-1)} \quad \text{for all } i \geq 1 \text{ and } k \geq 0.$$

Write down the above identities for $k = 1, \ldots, n$ and add them. This lead to the equation

$$\sum_{k=1}^{n} C_i^{k+i} = \sum_{k=0}^{n-1} C_i^{k+i} + \sum_{k=1}^{n} C_{i-1}^{k+(i-1)},$$

which can be rewritten as

$$C_i^{n+i} = C_0^{0+i} + \sum_{k=1}^{n} C_{i-1}^{k+(i-1)} = \sum_{k=0}^{n} C_{i-1}^{k+(i-1)}.$$

This proves (a)

(b) In view of the definition of the Fibonacci numbers of higher order, the following relations hold for given positive integers n and i:

$$f_n^{(i)} = f_n^{(i-1)} + f_{n-1}^{(i)} + f_{n-2}^{(i)}$$

$$f_{n-1}^{(i)} = f_{n-1}^{(i-1)} + f_{n-2}^{(i)} + f_{n-3}^{(i)}$$

$$2f_{n-2}^{(i)} = 2f_{n-2}^{(i-1)} + 2f_{n-3}^{(i)} + 2f_{n-4}^{(i)}$$

$$3f_{n-3}^{(i)} = 3f_{n-3}^{(i-1)} + 3f_{n-4}^{(i)} + 3f_{n-5}^{(i)}$$

$$5f_{n-4}^{(i)} = 5f_{n-4}^{(i-1)} + 5f_{n-5}^{(i)} + 5f_{n-6}^{(i)}$$

$$\dots \dots \dots \dots \dots \dots \dots \dots \dots$$

$$f_{n-2}^{(0)} f_2^{(i)} = f_{n-2}^{(0)} f_2^{(i-1)} + f_{n-2}^{(0)} f_1^{(i)} + f_{n-2}^{(0)} f_0^{(i)}$$

$$f_{n-1}^{(0)} f_1^{(i)} = f_{n-1}^{(0)} f_1^{(i-1)} + f_{n-1}^{(0)} f_0^{(i)}$$

$$f_n^{(0)} f_0^{(i)} = f_n^{(0)} f_0^{(i-1)}$$

By adding these identities we see that

$$f_n^{(i)} = f_0^{(0)} f_n^{(i-1)} + f_1^{(0)} f_{n-1}^{(i-1)} + \dots + f_n^{(0)} f_0^{(i-1)} \tag{2.16}$$

(c) According to (2.16)

$$f_{n-k}^{(i-1)} = \sum_{j=0}^{n-k} f_j^{(0)} f_{n-k-j}^{(i-2)} \quad \text{for} \quad k = 0, 1, \dots, n \tag{2.17}$$

Using (2.17), the formula for $f_n^{(i)}$ can be rewritten as follows:

$$f_n^{(i)} = \sum_{k=0}^{n} f_k^{(0)} f_{n-k}^{(i-1)} = \sum_{k=0}^{n} f_k^{(0)} \left(\sum_{j=0}^{n-k} f_j^{(0)} f_{n-k-j}^{(i-2)} \right). \tag{2.18}$$

By expanding the brackets in (2.18), $f_n^{(i)}$ can be expressed as a sum of products which consist of three factors: two Fibonacci numbers and a Fibonacci number of order $i - 2$.

If in the latter expansion each factor $f_\ell^{(i-2)}$ is replaced by $\sum_{j=0}^{\ell} f_j^{(0)} f_{\ell-j}^{(i-3)}$, then $f_n^{(i)}$ can be rewritten as the sum of products of four factors each: three Fibonacci numbers and a Fibonacci number of order $i - 3$.

By iterating the above procedure, we can express $f_n^{(i)}$ as a sum of products of $i + 1$ factors, each of which is a Fibonacci number.

Problem 18

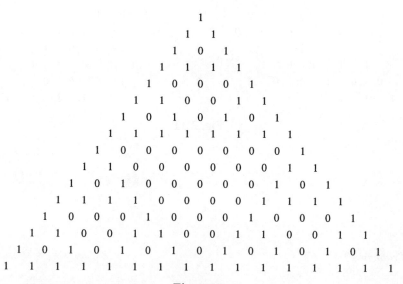

Fig. 2.10

Problem 19

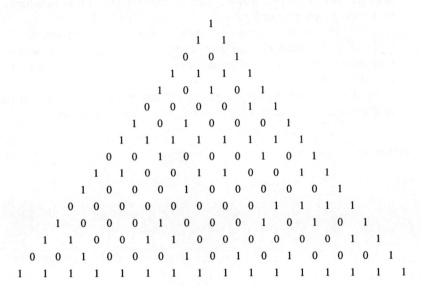

Fig. 2.11

Problem 20

(a)

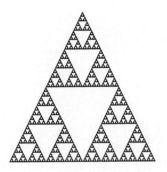

Fig. 2.12

(b)

Fig. 2.13

Problem 21

From

$$g_n(a, b) = ag_{n-1}(a, b) + bg_{n-2}(a, b)$$

and

$$h_n(a, b) = ah_{n-1}(a, b) + bh_{n-2}(a, b)$$

it follows that

$$Ag_n + Bh_n = A\left[ag_{n-1}(a, b) + bg_{n-2}(a, b)\right]$$

$$+ B\left[ah_{n-1}(a, b) + bh_{n-2}(a, b)\right]$$

$$= a\left[Ag_{n-1}(a, b) + Bh_{n-1}(a, b)\right]$$

$$+ b\left[Ag_{n-2}(a, b) + Bh_{n-2}(a, b)\right].$$

Problem 22

(a) Since α is a solution of the equation $x^2 = ax + b$,

$$\alpha^2 = a\alpha + b,$$

that is

$$\alpha^2 = a\alpha + b\alpha^0.$$

For any $n \geqslant 2$

$$\alpha^n = \alpha^{n-2} \cdot \alpha^2 = \alpha^{n-2}(a\alpha + b\alpha^0)$$

$$= a\alpha^{n-1} + b\alpha^{n-2}.$$

This proves (a).

(b) If α and β are distinct roots of the quadratic equation $x^2 = ax + b$ then, according to (a), both number sequences $\alpha^0, \alpha^1, \ldots, \alpha^n, \ldots$ and $\beta^0, \beta^1, \ldots, \beta^n, \ldots$ satisfy the recurrence relation $f_n(a, b) = af_{n-1}(a, b) + bf_{n-2}(a, b)$. Hence, from the statement of Problem 21, it follows that

$$A\alpha^0 + B\beta^0, A\alpha + B\beta, \ldots, A\alpha^n + B\beta^n, \ldots$$

satisfy the recurrence relation

$$A\alpha^n + B\beta^n = a(A\alpha^{n-1} + B\beta^{n-1}) + b(A\alpha^{n-2} + B\beta^{n-2}).$$

(c) The quadratic equation $x^2 - ax - b = 0$ has equal roots $\alpha = \beta$. Hence $a = 2\alpha$ and $-b = \alpha^2$.
In view of (a),

$$\alpha^n = a\alpha^{n-1} + b\alpha^{n-2}.$$

Thus

$$A\alpha^n + nB\alpha^n = A(a\alpha^{n-1} + b\alpha^{n-2}) + nB(a\alpha^{n-1} + b\alpha^{n-2})$$

$$= a[A\alpha^{n-1} + (n-1)B\alpha^{n-1}] + b[A\alpha^{n-2} + (n-2)B\alpha^{n-2}]$$

$$+ (aB\alpha^{n-1} + 2bB\alpha^{n-2}).$$

From $a = 2\alpha$ and $b = -\alpha^2$ it follows that

$$aB\alpha^{n-1} + 2bB\alpha^{n-2} = B\alpha^{n-2}[2\alpha \cdot \alpha + 2(-\alpha^2)] = 0.$$

Therefore the sequence $A, A\alpha + B\alpha, A\alpha^2 + 2B\alpha^2, \ldots, A\alpha^n + nB\alpha^n$ satisfies the recurrence relation $f_n(a, b) = af_{n-1}(a, b) + bf_{n-2}(a, b)$.

Problem 23

(a) The Fibonacci numbers $f_0, f_1, \ldots, f_n, \ldots$ satisfy the recurrence relation

$$f_n = af_{n-1} + bf_{n-2} \quad \text{for} \quad a = b = 1, \ n \geqslant 2.$$

The quadratic equation $x^2 = x + 1$ has two different solutions:

$$\alpha = \frac{1 + \sqrt{5}}{2} \quad \text{and} \quad \beta = \frac{1 - \sqrt{5}}{2}.$$

Thus

$$f_n = A\alpha^n + B\alpha^n \quad \text{for} \quad n \geqslant 0.$$

In particular,

$$1 = f_0 = A + B$$

and

$$1 = f_1 = A\alpha + B\beta$$

From the above system of simultaneous equations we obtain the values

$$A = \frac{1}{\sqrt{5}} \cdot \frac{1 + \sqrt{5}}{2} \quad \text{and} \quad B = \frac{1}{\sqrt{5}} \cdot \frac{1 - \sqrt{5}}{2}$$

Hence

$$f_n = \frac{1}{\sqrt{5}} \left[\left(\frac{1 + \sqrt{5}}{2} \right)^{n+1} - \left(\frac{1 - \sqrt{5}}{2} \right)^{n+1} \right].$$

This is the Binet formula.

(b) (i) $\bar{d}_n = 2\bar{d}_{n-1}$ for $n \geqslant 1$, and $\bar{d}_0 = 1$.

In this case it can be proved by straightforward induction that $\bar{d}_n = 2^n$ for $n \geqslant 0$.

(ii) $\bar{d}'_n = 2\bar{d}'_{n-1} + \bar{d}'_{n-2}$ for $n \geqslant 2$.

The corresponding quadratic equation $x^2 = 2x + 1$ has two distinct solutions: $\alpha = 1 + \sqrt{2}$ and $\beta = 1 - \sqrt{2}$. $\bar{d}'_0 = 1$ and $\bar{d}'_1 = 2$; therefore

$$A + B = 1$$

and

$$A(1 + \sqrt{2}) + B(1 - \sqrt{2}) = 2.$$

Hence

$$A = \frac{1}{2\sqrt{2}} (1 + \sqrt{2}), \quad B = -\frac{1}{2\sqrt{2}} (1 - \sqrt{2})$$

and

$$\bar{d}_n = \frac{1}{2\sqrt{2}} \left[(1 + \sqrt{2})^{n+1} - (1 - \sqrt{2})^{n+1} \right] \quad \text{for} \quad n \geqslant 0.$$

(iii) The numbers $\bar{s}_1, \bar{s}_2, \ldots, \bar{s}_n, \ldots$ are the Fibonacci numbers $f_0, f_1, \ldots, f_n, \ldots$. They were considered in case (a).

(iv) $\bar{s}'_n = \bar{s}'_{n-1} + 2\bar{s}'_{n-2}$ for $n \geqslant 2$. The solutions of the corresponding quadratic equation $x^2 = x + 2$ are $\alpha = 2$ and $B = -1$. The initial terms of the sequence are $s'_0 = \bar{s}'_1 = 1$; therefore

$$A + B = 1$$
$$2A - B = 1.$$

Hence $A = \dfrac{2}{3}$ and $B = \dfrac{1}{3}$, and

$$\bar{s}'_n = \frac{2}{3} \cdot 2^n + \frac{1}{3}(-1)^n \quad \text{for} \quad n \geqslant 0.$$

Chapter II

Problem 24

In order to find a method for solving this problem let us experiment by dividing the heaps of pebbles in two different ways, following the given rules, and compare the corresponding sums.

Experiment 1. At each stage divide each heap into two heaps as evenly as possible (see Fig. 2.14(a)). This yields the products $\pi_1 = 5 \cdot 6$, $\pi_2 = 3 \cdot 2$, $\pi_3 = 3 \cdot 3$, $\pi_4 = 2 \cdot 1$, $\pi_5 = 1 \cdot 1$, $\pi_6 = 2 \cdot 1$, $\pi_7 = 2 \cdot 1$, and $\pi_8 = \pi_9 = \pi_{10} = 1 \cdot 1$. Their sum $s = 55$.

Experiment 2. At each stage put only one pebble into one of the two newly created heaps (Fig. 2.14(b)). This process gives the products $\pi_i = (11 - i) \cdot 1 = 11 - i$ for $i - 1, 2, \ldots, 10$. In this case the sum s is again 55.

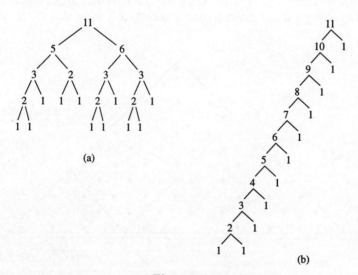

(a)

(b)

Fig. 2.14

151

Experiments 1 and 2 suggest the following *hypothesis.*

> The sum of the products $\pi_1, \pi_2, \pi_3, \ldots$ is always 55; it does not depend on the way in which, at each stage of the game, heaps with more than one pebble are divided into two non-empty sub-heaps.

It is a manageable, but a time-consuming, task to list all possible ways of forming sub-heaps at each stage of the puzzle and to calculate, in each case, the sum of the corresponding products. All sums obtained turn out to be equal to 55. A geometrical and an algebraic proof of this fact are described below in the solution of Problem 25.

Problem 25

(a) According to the solution of Problem 24, the sums of the products $\pi_1 = x(11 - x)$, $\pi_2 = y(x - y)$, $\pi_3 = z(11 - x - z)$, ... is 55, the 10th triangular number t_{10} (see p. 21). $t_{10} + 11 = t_{11}$, that is

$$\pi_1 + \pi_2 + \pi_3 + \ldots + 11 = t_{11} \qquad (2.19)$$

Triangular numbers can be represented by triangular arrays of dots. This suggests the idea of proving (2.19) geometrically by decomposing the triangular array of dots T_{11} that represents t_{11} into subsets of $\pi_1, \pi_2, \pi_3, \ldots$ and 11 dots. This can be done as follows:

Figure 2.15 shows the triangular array $T_{11} = ABC$, consisting of rows of $1, 2, 3, \ldots, 11$ dots. Mark the point D on CB such that the 11 dots along CB are divided into a subset of x dots along CD, and of

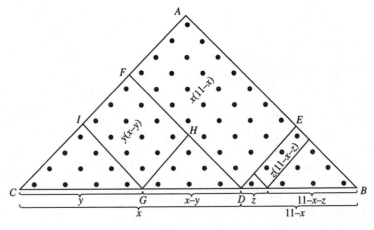

Fig. 2.15

$11 - x$ dots along DB. Through D draw parallels to CA and BA intersecting AB and AC at E and F respectively. In this way triangle ABC is divided into three parts:

the parallelogram $AEDF$ containing $x(11 - x)$ dots and thus representing the product π_1; triangle $FCD = T_x$, representing the triangular number T_x; triangle $EDB = T_{11-x}$, representing the triangular number t_{11-x}.

Hence

$$t_{11} = \pi_1 + t_x + t_{11-x} \tag{2.20}$$

for any x such that $1 \leqslant x < 11$.

The above procedure can be repeated; it can be applied to T_x if $x > 1$ and to T_{11-x} if $11 - x > 1$. Figure 2.15 depicts the partition of FCD into the parallelogram $FHGI$ with $y(x - y)$ dots representing π_2 and into triangles $IGC = T_y$ and $HDG = T_{x-y}$ representing t_y and t_{x-y} respectively. Thus

$$t_x = \pi_2 + t_y + t_{x-y}. \tag{2.21}$$

Similarly,

$$t_{11-x} = \pi_3 + t_z + t_{11-x-z}. \tag{2.22}$$

In view of (2.21) and (2.22), the identity (2.20) is transformed into

$$t_{11} = \pi_1 + \pi_2 + \pi_3 + t_y + t_{x-y} + t_z + t_{11-x-y}. \tag{2.23}$$

By iterating this procedure until each heap contains only one pebble, (2.23) is replaced by

$$t_{11} = \underbrace{\pi_1 + \pi_2 + \ldots}_{s} + \underbrace{1 + 1 + \ldots + 1}_{11}$$

This geometrical interpretation shows that $s = t_{10}$.

(b) Problem 24 can be extended to a heap of n pebbles as follows.
Divide a heap of n pebbles into two non-empty heaps containing x and $n - x$ pebbles respectively, and form the product $\pi_1 = x(n - x)$. Continue this procedure, that is at each stage divide each of the newly obtained heaps into two non-empty heaps (if possible) and record the corresponding product π_i. The process ends with n heaps of one pebble each. Find out whether the sum s_n of the products $\pi_1, \pi_2, \ldots$ depends on the manner of partitioning the original heap, and of each newly obtained 'sub-heap' into two non-empty sub-heaps.

Solution. The solution of the special case, $n = 11$, suggests that for any $n \geqslant 2$ the sum of the products $\pi_1, \pi_2, \ldots$ depends only on n, and not on the way in which the heap and the 'sub-heaps' are partitioned into two non-empty sub-heaps. Geometrical arguments, analogous to those illustrated in Fig. 2.15 can be used to prove that

$$s_n = t_{n-1} = \frac{n(n-1)}{2} \tag{2.24}$$

We now sketch a proof of (2.24) using mathematical induction.

If the original heap contains two pebbles, then $\pi_1 = 1 \cdot 1$ and $s_2 = 1$; this is equal to $t_1 = 2 \cdot 1/2$. Thus (2.24) is true for $n = 2$

Assume that (2.24) is true for all $n = 2, 3, .., k$ and consider a heap of $k + 1$ pebbles. Divide this heap into two non-empty heaps of x and $(k + 1) - x$ pebbles respectively. Then

$$s_{k+1} = x[(k+1) - x] + s_x + s_{(k+1)-x}.$$

If $x \geqslant 2$ then, since $x < k + 1$, the induction hypothesis implies that $s_x = t_{x-1}$. Similarly, if $k + 1 - x \geqslant 2$, then $s_{k+1-x} = t_{k-x}$. Thus

$$s_{k+1} = x(k+1-x) + \frac{x(x-1)}{2} + \frac{(k+1-x)(k-x)}{2}$$

$$= \frac{(k+1)k}{2} = t_k.$$

Hence (2.24) is true for all $n \geqslant 2$.

Problem 26

The solution of Problem 26 can be easily reduced to the solution of Problem 25. In fact, the division of a heap of n pebbles into three heaps of x, y and $n - (x + y)$ pebbles respectively can be carried out in two steps: the heap of n pebbles is first divided into two heaps of x and $n - x$ pebbles, and then the latter heap is divided into two heaps containing y and $(n - x) - y = n - (x + y)$ pebbles. The expression ε_1, corresponding to the division of the original heap into three parts, is equal to $xy + x(n - x - y) + y(n - x - y)$. This can be rewritten as

$$\underbrace{x(n-x)}_{\pi_1} + \underbrace{y[(n-x)-y]}_{\pi_2} ,$$

where π_1 and π_2 are the products defined in Problem 25. Proceeding further, the division of any sub-heap into three non-empty parts can be replaced by two divisions of sub-heaps into two non-empty parts each. Thus, in view of the solution of Problem 25,

$$\varepsilon_1 + \varepsilon_2 + \ldots = \pi_1 + \pi_2 + \ldots = t_{n-1} \quad \text{for all} \quad n \geqslant 3 \tag{2.25}$$

Figure 2.16 provides a geometrical interpretation of the solution. Triangle $ABC = T_n$, the triangular pattern of dots representing the nth triangular number t_n. The points D and E on the side CB are taken so that there are x points on CD, y points on DE, and $n - (x + y)$ points on EB. Then two points I and H are chosen on AC so that there are x points on CI, $n - (x + y)$ points on IH, and y points on HA. Finally, two points, G and F, are marked on AB so that there are y points on AG, x points on GF, and $n - (x + y)$ points on FB. Thus T_n is divided into four parts: three triangles T_x, T_y, and T_{n-x+y}, and a hexagonal pattern of dots with perimeter $DE + EF + FG + GH + HI + ID$. The hexagon $DEFGHI$ is partitioned into three parallelograms: $HGFJ$ with xy dots, $JFED$ with $y(n - x - y)$ dots, and $HJDI$ with $x(n - x - y)$ dots. Hence

$$T_n = T_x + T_y + T_{n-x-y} + \varepsilon_1.$$

By iterating this process, that is by dividing at each stage the patterns T_k into three triangles and three parallelograms, we find that

$$\varepsilon_1 + \varepsilon_2 + \ldots = t_{n-1}.$$

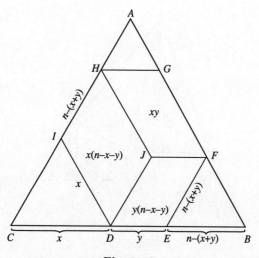

Fig. 2.16

Problem 27

The partition of a heap of n pebbles into two non-empty sub-heaps corresponds to the division of a straight-line segment of length n into two line segments of length x and y. The expression $\zeta_1 = x^2y + xy^2$ can be thought of as a sum of volumes of solids with some edges of length x and y. Figure 2.17 shows such a geometrical interpretation.

$P_n = EABCD$ is a pyramid of altitude $ED = n$ and square base $ABCD$ of side length $AB = n$. Points F, G, and H are marked on the edges DA, DC, and DE respectively, such that $DF = DG = x$ and $DH = y$. Hence $FA = GC = n - x = y$ and $HE = n - y = x$.

Now we construct three planes: a plane through H, parallel to $ABCD$ and intersecting P in a square $HKJI$; a plane through G, parallel to EDA and intersecting P in a trapezoid $GMJK$. The trapezoids $FNJI$ and $GMJK$ meet along a line segment JL.

In this way P_n is divided into five parts: the cuboid $FLGDIJKH$, of volume x^2y, the triangular prisms $AFIMLJ$ and $NLJCGK$ of volume $\frac{1}{2} y^2x$ each, and the pyramids $P_x = EIJKH$ and $P_y = JMBNL$.

Thus

$$\text{Volume } P_n = \zeta_1 + \text{Volume } P_x + \text{Volume } P_y.$$

At the next stage the pyramid P_x is partitioned into a cuboid of volume u^2z (where $u + z = x$), two triangular prisms of volume $\frac{1}{2} uz^2$ each, and two pyramids P_u and P_z; pyramid P_u is partitioned in a similar way.

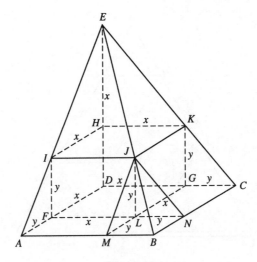

Fig. 2.17

Iteration of the above process stops when all the pyramids resulting from the partitions have square bases of side length 1, and hence altitudes of length 1. At the final stage there will be n such pyramids. Hence

$$\zeta_1 + \zeta_2 \ldots = \text{Volume } P_n - n \cdot \frac{1^2 \cdot 1}{3}$$

$$= \frac{n^3}{3} - \frac{n}{3} = \frac{n(n^2 - 1)}{3}. \tag{2.26}$$

Problem 28

(a) Figure 2.18 shows the first few steps in the process of partitioning heaps of f_i pebbles into two heaps with f_{i-1} and f_{i-2} pebbles, $i = k, k - 1, k - 2, \ldots$. These steps lead to the first products π_i defined in Problem 25:

$$\pi_1 = f_{k-1}f_{k-2}, \quad \pi_2 = f_{k-2}f_{k-3}, \quad \pi_3 = \pi_4 = f_{k-3}f_{k-4},$$

$$\pi_5 = \pi_6 = \pi_7 = f_{k-4}f_{k-5}.$$

The diagram suggests that the partitions in Problem 28 lead to the following special case of (2.24)

$$t_{f_{k-1}} = \sum_{i=0}^{k-2} f_i f_{k-(i+1)} f_{k-(i+2)}.$$

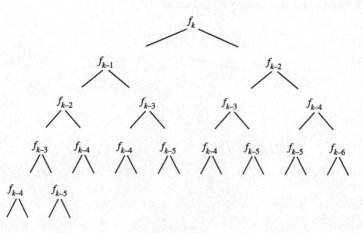

Fig. 2.18

Thus

$$\frac{f_k(f_k-1)}{2} = \sum_{i=0}^{k-2} f_i f_{k-(i+1)} f_{k-(i+2)}.$$

It is left to the reader to prove this formula for all Fibonacci numbers $f_k, k \geqslant 2$, using mathematical induction.

(b) According to Fig. 2.18, the first few summands ζ_i of s defined in Problem 27 are $\zeta_1 = f_{k-1}^2 f_{k-2} + f_{k-1} f_{k-2}^2$, $\zeta_2 = f_{k-2}^2 f_{k-3} + f_{k-2} f_{k-3}^2$, $\zeta_3 = \zeta_4 = f_{k-3}^2 f_{k-4} + f_{k-3} f_{k-4}^2, \dots$. This suggests that in our special case formula (2.26) can be rewritten as

$$\frac{f_k(f_k^2-1)}{3} = \sum_{i=0}^{k-2} f_i \left(f_{k-(i+1)}^2 f_{k-(i+2)} + f_{k-(i+1)} f_{k-(i+2)}^2 \right).$$

Again, it is left to the reader to verify this formula.

Problem 29

Just as in the previous problem, Fig. 2.19 suggests the following special case of (2.25):

$$\frac{f_k(3)[f_k(3)-1]}{2} = \sum_{i=0}^{k-3} f_i(3)\big[f_{k-(i+1)}(3) f_{k-(i+2)}(3) + f_{k-(i+1)}(3) f_{k-(i+3)}$$

$$+ f_{k-(i+2)}(3) f_{k-(i+3)}(3)\big]$$

This formula can be verified by induction.

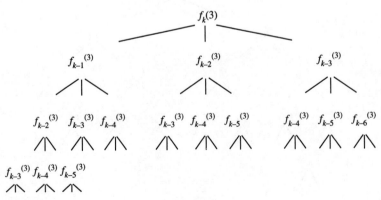

Fig. 2.19

Problem 30

We establish (a) and (b) simultaneously. A regular $2n$-gon can be assembled from t_{n-1} rhombuses as follows. From a point $A^{(0)}$ drawn n congruent straight-line segments $A^{(0)} A_1^{(1)}, A^{(0)} A_2^{(1)}, \ldots, A^{(0)} A_n^{(1)}$ such that

$$\angle A_j^{(1)} A^{(0)} A_{j+1}^{(1)} = \alpha = \frac{180°}{n} \quad \text{for} \quad j = 1, 2, \ldots, n-1.$$

Construct points $A_j^{(2)}$ for $j = 1, 2, \ldots, n-1$ such that the quadrilaterals $R_j^{(1)} = A^{(0)} A_j^{(1)} A_j^{(2)} A_{j+1}^{(1)}$ are rhombuses. The angles of the rhombuses $R_j^{(1)}$ are α and $180° - \alpha$.

Proceed by constructing points $A_j^{(3)}$ for $j = 1, \ldots, n-2$, points $A_j^{(4)}$ for $j = 1, \ldots, n-3, \ldots,$ $A_j^{(i)}$ for $j = 1, \ldots, n-(i-1), \ldots, A_1^{(n)}$ such that the quadrilaterals $R_j^{(i)} = A_{j+1}^{(i-2)} A_j^{(i-1)} A_j^{(i)} A_{j+1}^{(i-1)}$ are rhombuses for $i = 2, 3, \ldots, n-1$ and $j = 1, 2, \ldots, n-i$ (Fig. 2.20). The angles of $R_j^{(i)}$ are $i\alpha$ and $180° - i\alpha$.

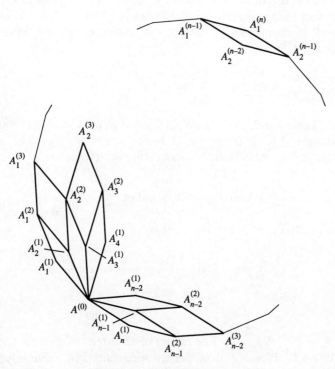

Fig. 2.20

In this way we obtain a 2n-gon $P_{2n} = A^{(0)} A_1^{(1)} A_1^{(2)} \ldots A_1^{(n)} A_2^{(n-1)} A_3^{(n-2)} \ldots$ $A_{n-1}^{(2)} A_n^{(1)}$. The sides of P_{2n} are all congruent to $A^0 A_1^{(1)}$. The angles of P_{2n} are $(n-1)180°/n$. This is so because

$$\angle A_1^{(1)} A^{(0)} A_n^{(1)} = (n-1)\alpha, \qquad \angle A_1^{(i+1)} A_1^{(i)} A_1^{(i-1)} = \angle A_{n-i}^{(i+1)} A_{n+1-i}^{(i)} A_{n+2-i}^{(i-1)}$$

$$= (180° - i\alpha) + (i-1)\alpha = 180° - \alpha = n\alpha - \alpha = (n-1)\alpha$$

$$\text{for} \quad i = 1, \ldots, n-1$$

and

$$\angle A_1^{(n-1)} A_1^{(n)} A_2^{(n-1)} = (n-1)\alpha.$$

P_{2n} is a regular $2n$-gon. It is partitioned into t_{n-1} rhombuses: $(n-1)$ rhombuses with angles α and $180° - \alpha$, $(n-2)$ rhombuses with angles 2α and $180° - 2\alpha, \ldots$, two rhombuses with angles $(n-2)\alpha$ and $180° - (n-2)\alpha$, and one rhombus with angles $(n-1)\alpha$ and $180° - (n-1)\alpha$.

All regular $2n$-gons are similar. Therefore any regular $2n$-gon can be divided into t_{n-1} rhombuses in the same way as P_{2n}.

(c) Let a be the length of the sides of P_{2n}. The area of P_{2n} can be expressed as the sum of the areas of the rhombuses $R_j^{(i)}$. For $i = 1, 2, \ldots, n-1$ the areas of the rhombuses $R_1^{(i)}, R_2^{(i)}, \ldots, R_{n-i}^{(i)}$ are $a^2\sin(i\alpha) = a^2\sin i \cdot 180°/n$ respectively. Hence

$$\text{Area } P_{2n} = \sum_{i=1}^{n-1} (n-i) a^2 \sin \frac{i \cdot 180°}{n}. \qquad (2.27)$$

On the other hand, P_{2n} can be partitioned into $2n$ triangles, congruent to triangle $A^{(0)} A_1^{(1)} O$, where O is the centre of P_{2n}. The area of triangle $A^{(0)} A_1^{(1)} O$ is $\frac{1}{2}a(a/2) \cotan (360°/4n)$. Hence

$$\text{Area } P_{2n} = 2n\frac{a^2}{4} \cotan \frac{90°}{n}. \qquad (2.28)$$

From (2.27) and (2.28) it follows that

$$\sum_{i=1}^{n-1} (n-i) \sin \frac{i \cdot 180°}{n} = \frac{n}{2} \cotan \frac{90°}{n}.$$

Problem 31

(a) Let $A_k = \{P_1, P_2, \ldots, P_{s_k}\}$ be the set of partitions of n into k parts, and let $B_k = \{P_1', P_2', \ldots, P_{t_k}'\}$ be the set of partitions of n into integers the largest of which is k. Our aim is to prove that $s_k = t_k$.

Each partition $P_i \in A$ is represented by a Ferrer graph F_i and each partition $P_j' \in B$ is represented by a Ferrer graph F_j'. Consider the sets $\mathscr{F}_k' = \{F_1, F_2, \ldots, F_{s_k}\}$ and $\mathscr{F}_k' = \{F_1', F_2', \ldots, F_{t_k}'\}$. On the set $\mathscr{F}_k$ define a transformation α as follows.

$\alpha(F_i)$ is the diagram obtained by a rotation of F_i through $90°$ anti-clockwise, followed by a reflection of the rotated image in a vertical axis (see Fig. 1.19).

The transformation α has the following properties:

(1) $\alpha(F_i) \in \mathscr{F}_k'$ for all $i = 1, \ldots, s_k$;

(2) $\alpha(F_i) = \alpha(F_j)$ implies that $i = j$ for $i, j = 1, 2, \ldots, s_k$;

(3) for each $F_j' \in \mathscr{F}_k'$ there exists $F_i \in \mathscr{F}_k$ such that $F_j' = \alpha(F_i)$.

Properties (1)–(3) imply that α establishes a one-to-one correspondence between the elements of $\mathscr{F}_k$ and $\mathscr{F}_k'$. Hence $s_k = t_k$.

(b) Denote by $\overline{A}_k$ the set of partitions of n into at most k parts and by $\overline{B}_k$ the set of partitions of n into parts which do not exceed k. The sets $\overline{A}_k$ and $\overline{B}_k$ are the disjoint unions of the following subsets:

$$\overline{A}_k = A_1 \cup A_2 \cup \ldots \cup A_k$$

and

$$\overline{B}_k = B_1 \cup B_2 \cup \ldots \cup B_k$$

According to (a), $s_i = t_i$ for $i = 1, \ldots, k$. Since the sets A_i as well as the sets B_i are pairwise disjoint, $s_1 + s_2 + \ldots + s_k = t_1 + t_2 + \ldots + t_k$, that is $\overline{A}_k$ and $\overline{B}_k$ contain the same number of elements.

Problem 32

(a) This part of the problem will be solved in several steps. Divide the set of S of all partitions of n into the subset A of those partitions whose Ferrer graphs consist of more than one block, and the subset B of the remaining partitions. In the Ferrer graph of any partition of n denote the number of dots in the top row by p_1, and the number of rows in the bottom block by i.

Step. 1. Consider first the partitions of A and their corresponding Ferrer graphs. Two cases can be distinguished.

Case 1. $p_1 \leqslant i$. In this case the Ferrer graph admits the transformation α (and not β). Denote the subset of A consisting of the partitions with such Ferrer graphs by A_α.

Case 2. $p_1 > i$. Then the Ferrer graph admits β (and not α). Denote the subset of A which consists of partitions with this type of Ferrer graphs by A_β.

The conditions $p_1 \leqslant i$ and $p_i > i$ cover all possible relations between p_1 and i. Hence

$$A = A_\alpha \cup A_\beta, \quad \text{and} \quad A_\alpha \cap A_\beta = \emptyset.$$

Step 2. Now consider the partitions of B with their Ferrer graphs. Several cases can be distinguished.

Case 1. $p_1 \leqslant i - 1$. In this case the Ferrer graph admits α (and not β). Denote the subset of B consisting of partitions with such Ferrer graphs by B_α.

Case 2. $p_1 \geqslant i + 2$. In this case the Ferrer graph admits β (and not α). Denote the subset of B which consists of partitions with such Ferrer graphs by B_β.

Case 3. $p_1 = i$. In this case the Ferrer graph has a single block with i rows, containing $i, i + 1, i + 2, \ldots, i + (i - 1)$ dots respectively. Thus $n = i + (i + 1) + (i + 2) + \ldots + [i + (i - 1)] = (3i - 1)i/2$. This implies that if n cannot be expressed as $(3i^2 - i)/2$ for some natural number i, then Case 3 cannot occur.

Case 4. $p_1 = i + 1$. In this case the single block of the corresponding Ferrer graph has i rows with $i + 1, i + 2, \ldots, i + 1 + (i - 1)$ dots respectively. Hence $n = (i + 1) + (i + 2) + \ldots + [i + 1 + (i - 1)] = (3i + 1)i/2$. If n cannot be expressed as $(3i^2 + i)/2$ for some natural number i, then case 4 cannot occur.

It now follows that if n is not equal to $(3i^2 - i)/2$ or $(3i^2 + i)/2$ for any natural number i, then the set S of all partitions of n is the union of the following, pairwise disjoint subsets:

$$S = A_\alpha \cup A_\beta \cup B_\alpha \cup B_\beta = (A_\alpha \cup B_\alpha) \cup (A_\beta \cup B_\beta).$$

Every partition of n belongs either to $A_\alpha \cup B_\alpha$ or to $A_\beta \cup B_\beta$; hence its Ferrer graph admits either α or β.

(b) If $n = (3i^2 - i)/2$, then $S = (A_\alpha \cup B_\alpha) \cup (A_\beta \cup B_\beta) \cup \{\overline{P}\}$, where $\overline{P}$ is the partition of n into summands $i, i + 1, \ldots, i + (i - 1)$. $\overline{P}$ admits neither α nor β.

If $n = (3i^2 + i)/2$, then $S = (A_\alpha \cup B_\alpha) \cup (A_\beta \cup B_\beta) \cup \{\overline{P}'\}$, where $\overline{P}'$ is the partition of n into summands $i + 1, i + 2, \ldots, i + 1 + (i - 1)$. $\overline{P}'$ admits neither α nor β.

Problem 33

(a) For any partition P of n denote by F_P its corresponding Ferrer graph. The discussion in Problem 32 implies that:

   (1) If $P \in A_\alpha \cup B_\alpha$ then F_P admits α and $\alpha(F_P) \in A_\beta \cup B_\beta$.

   (2) If $P \in A_\beta \cup B_\beta$ then F_P admits β and $\beta(F_P) \in A_\alpha \cup B_\alpha$.

The definitions of α and β imply that

   (3) If $P \in A_\alpha \cup B_\alpha$ then $\beta(\alpha(F_P)) = F_P$.

From (1)–(3) it is easy to deduce that

   (4) α is a one-to-one mapping of the set $\{F_P : P \in A_\alpha \cup B_\alpha\}$ onto the set $\{F_P : P \in A_\beta \cup B_\beta\}$.

For any $P \in A_\alpha \cup B_\alpha$ the number of rows in the Ferrer diagram of $\alpha(F_P)$ is one less than the number of rows in F_P. Thus

   (5) α maps any 'even' Ferrer diagram (a diagram with an even number of rows) onto an 'odd' Ferrer diagram (with an odd number of rows) and conversely.

From (4) and (5) it follows that if $\{F_P : P \in A_\alpha \cup B_\alpha\}$ contains x even and y odd diagrams, then $\{F_P : P \in A_\beta \cup B_\beta\}$ contains x odd and y even diagrams. Thus the union of these disjoint sets consists of $x + y$ even and $x + y$ odd diagrams.

If $n \neq (3i^2 - i)/2$, $(3i^2 + i)/2$ for any natural number i then, in view of the solution of Problem 32(a), the set $S = (A_\alpha \cup B_\alpha) \cup (A_\beta \cup A_\beta)$. Hence $O_n = x + y = E_n$.

(b) In $n = (3i^2 - i)/2$ or $n = (3i^2 + i)/2$ for some natural number i then, according to the solution of Problem 32(b) (cases 3 and 4), there is a unique partition $\overline{P}$ of n not contained in $(A_\alpha \cup B_\alpha) \cup (A_\beta \cup B_\beta)$. The Ferrer diagram of this partition has i rows. Thus the partition $\overline{P}$ is even if i is even and odd otherwise. Hence

$$E_n - O_n = (x + y + 1) - (x + y) = 1 \text{ for } i \text{ even}$$

and

$$E_n - O_n = (x + y) - (x + y + 1) = -1 \text{ for } i \text{ odd}$$

In other words

If $n = \dfrac{3i^2 - i}{2}$ or $n = \dfrac{3i^2 + i}{2}$, then $E_n - O_n = (-1)^i$ for $i = 1, 2, \ldots$.

Problem 34

Let Π_n^* denote the result of multiplying the factors of $\Pi_n = (1-x)(1-x^2)\ldots(1-x^n)$ prior to collecting like terms.

In Π_n^*, for each natural number $m \leqslant n$, the terms x^m are formed by products $(-x^{p_1})(-x^{p_2})\ldots(-x^{p_k}) = (-1)^k x^{p_1+p_2+\ldots+p_k}$ for all possible partitions of m into distinct parts. More specifically, for each partition of m into distinct parts there is a corresponding term x^m in Π_n^* whose sign is positive if the partition is even and negative otherwise. Hence, after collecting like terms, Π_n^* is transformed into a sum starting with the following $n+1$ summands:

$$1 - x - x^2 + (E_3 - O_3)x^3 + (E_4 - O_4)x^4 + \ldots + (E_n - O_n)x^n.$$

Thus

$$\Pi_n = 1 - x - x^2 + (E_3 - O_3)x^3 + \ldots + (E_n - O_n)x^n + a_{n+1}x^{n+1}$$

$$+ a_{n+2}x^{n+2} + \ldots + a_{n(n+1)/2}x^{n(n+1)/2}$$

with some coefficients $a_{n+1}, a_{n+2}, \ldots, a_{n(n+1)/2}$.

Problem 35

Since

$$\Pi_n = 1 - x - x^2 + \sum_{k=3}^{n}(E_k - O_k)x^k + \sum_{k=n+1}^{n(n+1)/2} a_k x^k \quad \text{for all} \quad n \geqslant 3$$

it follows that

$$\lim_{n\to\infty} \Pi_n = 1 - x - x^2 + \lim_{n\to\infty}\sum_{k=3}^{n}(E_k - O_k)x^t,$$

that is

$$\prod_{n=1}^{\infty}(1-x^n) = 1 - x - x^2 + \sum_{k+3}^{\infty}(E_k - O_k)x^k. \qquad (2.29)$$

According to the solution of Problem 33, $E_k - O_k = 0$ if k is not expressible as $(3i^2 \pm i)/2$. If $k = (3i^2 \pm i)/2$ then $E_k - O_k = (-1)^i$. Hence (2.29) can be rewritten as

$$\Pi(1-x^n) = 1 - x - x^2 + \sum_{i=1}^{\infty}(-1)^i\left(x^{(3i^2-i)/2} + x^{(3i^2+i)/2}\right)$$

$$= 1 + \sum_{i=1}^{\infty}(-1)^i\left(x^{(3i^2-i)/2} + x^{(3i^2+i)/2}\right).$$

This finishes the proof of Euler's formula (1.17).

Chapter III

Problem 36

Let a/b be a reduced fraction such that $0 < a/b < 1$. First we shall prove the following.

(i) a/b has a *finite* number of digits in its decimal expansion if and only if its denominator b is of the form $2^\alpha \cdot 5^\beta$ for some non-negative integers α and β.

(a) Suppose that a/b is in reduced form and has a finite number of digits in its decimal expansion: $a/b = 0.a_1 a_2 \dots a_n$. Then

$$0.a_1 a_2 \dots a_n = \frac{\overline{a_1 a_2 \dots a_n}}{10^n},$$

where $\overline{a_1 a_2 \dots a_n}$ denotes the number with digits $a_1, a_2, \dots, a_n$. The only prime factors of 10^n are 2 and 5. Hence

$$\frac{a}{b} = \frac{a}{2^\alpha \cdot 5^\beta} \quad \text{for some} \quad \alpha \geqslant 0, \quad \beta \geqslant 0,$$

(b) Suppose that a/b is in reduced form and that $b = 2^\alpha \cdot 5^\beta$. Let $\gamma = \max(\alpha, \beta)$. In that case,

$$\frac{a}{b} = \frac{a \cdot 2^{\gamma-\alpha} \cdot 5^{\gamma-\beta}}{2^\gamma \cdot 5^\gamma} = \frac{\overline{a_1 a_2 \dots a_n}}{10^\gamma}.$$

The decimal expansion of the fraction $\overline{a_1 a_2 \dots a_n}/10^\gamma$ has γ (a finite number) decimal places. This proves (i).

(i) implies that

(ii) a/b has an *infinite* number of digits in its decimal expansion if and only if the denominator b has at least one prime factor different from 2 and 5.

Problem 37

Suppose that a/b is a reduced fraction less than 1 with an infinite decimal expansion. This expansion is obtained by the usual method of long division. The remainders in the division belong to the set $\{1, 2, \dots, b-1\}$

165

(0 cannot appear as a remainder because the expansion does not terminate). Hence in the sequence of successive remainders there must be a repetition at the kth step for some $k \leqslant b - 1$. This implies that the decimal expansion of a/b is periodic; the length of the period does not exceed $b - 1$.

Problem 38

Suppose that a/b is a reduced fraction less than 1 such that its denominator b is coprime to 10. According to the assertions in Problems 36 and 37, the fraction a/b has an infinite decimal periodic expansion. Thus in the process of long division that yields the decimal expansion of a/b we obtain equal remainders $r_m = r_n$ for some $m > n$.

In the process of long division the remainder r_i, obtained at the ith step for $i \geqslant 1$, arises from the division of $10r_{i-1}$ by b, where r_{i-1} is the remainder obtained at the previous, $(i - 1)$th step. Thus

$$\text{and} \quad \left. \begin{array}{l} 10r_{m-1} = bq_{m-1} + r_m \\ 10r_{n-1} = bq_{n-1} + r_n \end{array} \right\} \tag{2.30}$$

for some natural numbers q_{m-1} and q_{n-1}.

Since according to our assumption $r_m = r_n$, the identities (2.28) imply that

$$10(r_{m-1} - r_{n-1}) = b(q_{m-1} - q_{n-1}).$$

Hence b divides $10(r_{m-1} - r_{n-1})$. Since b is coprime to 10, it follows that b must divide $r_{m-1} - r_{n-1}$. Both r_{m-1} and r_{n-1} are less than b; therefore b can divide their difference $r_{m-1} - r_{n-1}$ only if $r_{m-1} - r_{n-1} = 0$, that is if $r_{m-1} = r_{n-1}$.

By similar arguments we show that $r_{m-i} = r_{n-i}$ for all stages $n - i$ in the process of the long division, starting with $r_{n-i} = r_1$. Thus the period of the decimal expansion of a/b starts immediately after the decimal point; in other words, the decimal expansion of a/b is purely periodic.

Problem 39

Let a/b be a reduced fraction less than 1 with a purely periodic decimal expansion. Denote the length of the period of a/b by L. In that case

$$\frac{a}{b} = \frac{a_1}{10} + \frac{a_2}{10^2} + \ldots + \frac{a_L}{10^L} + \frac{a}{10^L b} \quad \text{for some integers} \quad a_1, a_2, \ldots, a_L,$$

so that

$$10^L a = b(10^{L-1}a_1 + 10^{L-2}a_2 + \ldots + a_L) + a.$$

hence $10^L a - a = a(10^L - 1)$ is divisible by b. Since a and b are coprime, this implies that b divides $10^L - 1$.

The period in the decimal expansion of a/b ends after the *first* repetition of the remainder a in the process of long division. Thus L is the smallest natural number for which $10^L a$ and a leave the same remainder when divided by b. Therefore L is the smallest number for which $10^L - 1$ is divisible by b.

Problem 40

Since b is coprime to 2 and 5, the decimal expansion of the fraction $1/b$ is purely periodic. According to the statement proved in Problem 39, the period of this decimal expansion is the smallest number L for which $10^L - 1$ is divisible by b. The difference $10^L - 1$ is equal to the product

$$10^L - 1 = (10 - 1)(10^{L-1} + 10^{L-2} + \ldots + 10^2 + 10 + 1) = 9 \cdot \underbrace{11\ldots 1}_{L}.$$

Since b divides $10^L - 1$, and since b is coprime to 3, this implies that the repunit $\underbrace{111\ldots 1}_{L}$ is a multiple of b.

Problem 41

(a) The long division $1 \div 7$ produces the sequence of quotients and corresponding remainders shown in Table 1.

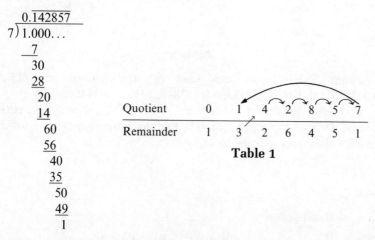

```
 0.142857
7) 1.000...
   7
   30
   28
   20
   14
   60
   56
   40
   35
   50
   49
    1
```

Quotient	0	1	4	2	8	5	7
Remainder	1	3	2	6	4	5	1

Table 1

If i is a remainder in the long division $1 \div 7$, then the period of $i/7$ can be constructed by a cyclic permutation of the period $\overline{142857}$, starting with the quotient $10i \div 7$. This construction is indicated by the arrows in Table 1 for $i = 3$:

$$\frac{3}{7} = 0.\overline{428571}.$$

Since 2, 4, 5, and 6 are also remainders in the long division $1 \div 7$, Table 1 tells us that

$$\frac{2}{7} = 0.\overline{285714}, \quad \frac{4}{7} = 0.\overline{571428}, \quad \frac{5}{7} = 0.\overline{714285}, \text{ and } \frac{6}{7} = 0.\overline{857142}.$$

(b) The long division $1 \div 13$ yields the sequence of quotients and corresponding remainders shown in Table 2.

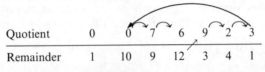

Quotient	0	0	7	6	9	2	3
Remainder	1	10	9	12	3	4	1

Table 2

Table 2 contains the remainders $i = 1, 10, 9, 12, 3, 4$. For these values of i the periods of the decimal expansions of $i/13$ are given in this table; for example, $12/13 = 0.\overline{923076}$.

Table 2 does not help us to find the decimal expansion of $2/13$; the long division $2 \div 13$ leads to a new table of remainders with corresponding quotients (Table 3).

Quotient	0	1	5	3	8	4	6
Remainder	2	7	5	11	6	8	2

Table 3

From Table 3 we can read off the periods of $i/13$ for $i = 2, 7, 5, 11, 6, 8$; e.g. $2/13 = 0.\overline{153846}$, $5/13 = 0.\overline{384615}$.

It is left to the reader to construct similar tables, showing the periods of the decimal expansions of $i/21$ for any natural number i which is smaller than and coprime to 21.

Problem 42

For any natural number n

$$\frac{1}{n} + \frac{1}{n+1} + \frac{1}{n+2} = \frac{3n^2 + 6n + 2}{n(n+1)(n-2)}.$$

Denote the reduced form of the fraction $(3n^2 + 6n + 2)/n(n+1)(n-2)$ by a/b. a/b has the following properties.

(1) b is divisible by 3. This follows from the fact that $3n^2 + 6n + 2 = 3(n^2 + 2n) + 2$ is *not* divisible by 3, while $n(n + 1)(n + 2)$, as the product of three consecutive natural numbers n, $n + 1$, and $n + 2$, is divisible by 3.

(2) b is divisible by 2. For: if n is odd then $3n^2 + 6n + 2$ is odd, and $n(n + 1)(n + 2)$ is even; if n is even then $3n^2 + 6n + 2$ is *not* divisible by 4, and $n(n + 1)(n + 2)$ is divisible by 4 (because both factors n and $n + 2$ are even).

In view of (1), the fraction a/b has an infinite decimal expansion (see Problem 36).

Suppose that the decimal expansion of a/b is purely periodic with period of length L. This implies that

$$\frac{a}{b} = \frac{a_1}{10} + \frac{a_2}{10^2} + \ldots + \frac{a_L}{10^L} + \frac{a}{10^L b}.$$

Hence

$$10^L a - a = b(10^{L-1}a_1 + 10^{L-2}a_2 + \ldots + a_L),$$

that is, b divides $10^L a - a = a(10^L - 1)$. Since b and a are coprime, b divides $10^L - 1$. However, according to (2), b is even; hence b cannot divide the odd number $10^L - 1$. This contradiction shows that the decimal expansion of a/b is not purely periodic.

Problem 43

Suppose that the decimal number $d = 0.12345678910111213\ldots$ is periodic, that its period starts after the kth decimal place and that the period is of length L.

The construction of d implies that $k + L + 1$ successive decimal places of d will be occupied by the digits of the number 10^{k+L}, that is by 1 followed by 10^{k+L} zeros. Hence the period of d is $\bar{0}$; this means that d is a finite decimal number.

This contradiction shows that a/b is not a periodic decimal fraction.

Problem 44

(a) The fraction a_n is in its lowest terms if and only if its reciprocal is in its lowest terms.

$$\frac{1}{a_n} = \frac{n^4 + 3n^2 + 1}{n^3 + 2n} = n + \frac{n^2 + 1}{n^3 + 2n} = n + a'_n.$$

$1/a_n$ is a fraction in its lowest terms if and only if a'_n is a fraction in its lowest terms, that is if and only if $1/a'_n$ is a fraction in its lowest terms.

$$\frac{1}{a'_n} = \frac{n^3 + 2n}{n^2 + 1} = 1 + \frac{n}{n^2 + 1}.$$

The numbers n and $n^2 + 1$ are coprime. Thus $1/a'_n$ is a fraction in its lowest terms.

From the above arguments it follows that a_n is a reduced fraction.

(b) The denominator $n^4 + 3n^2 + 1$ of a_n has the following properties:

(1) $n^4 + 3n^2 + 1$ is not divisible by 2 for $n = 1, 2, \ldots$;

(2) $n^4 + 3n^2 + 1$ is divisible by 5 if and only if $n = 5k \pm 1$ for $k = 1, 2, \ldots$;

(3) $n^4 + 3n^2 + 1$ has a prime factor different from 2 or 5 for $n = 2, 3, \ldots$

(1)–(3) can be proved as follows.

Proof of (1). $n^4 + 3n^2 + 1 = n^2(n^2 + 3) + 1$. If n is odd then $n^2 + 3$ is even, and therefore $n^2(n^2 + 3) + 1$ is odd. If n is even than n^2 is even, hence $n^2(n^2 + 3) + 1$ is odd.

Proof of (2). Suppose that $n^4 + 3n^2 + 1$ is divisible by 5. This implies that $(n^2 - 1)[(n^2 - 1) + 5] = 5m$ for some natural number m.

Hence

$$(n^2 - 1)^2 = 5m - 5(n^2 - 1) - 5.$$

Thus 5 divides $n^2 - 1$, that is

either 5 divides $n - 1$, in which case $n = 5k + 1$ for $k \in \{1, 2, \ldots\}$,
or 5 divides $n + 1$, in which case $n = 5k - 1$ for $k \in \{1, 2, \ldots\}$.

It is easy to verify the converse statement: If $n = 5k \pm 1$ for any $k = 1, 2, \ldots$, then $n^4 + 3n^2 + 1$ is divisible by 5.

Proof of (3). Suppose that $n^4 + 3n^2 + 1$ has no prime factor different from 2 or 5. Since $n^4 + 3n^2 + 1$ is odd, this implies that

$$n^4 + 3n^2 + 1 = 5^a \text{ for some natural number } a.$$

The last identity can be rewritten in the form

$$(n^2 - 1)[(n^2 - 1) + 5] = 5(5^{a-1} - 1).$$

Thus 5 divides the product $(n^2 - 1)[(n^2 - 1) + 5]$, that is, it divides at least one of the factors $n^2 - 1$ and $(n^2 - 1) + 5$. However, these factors differ by 5; hence both $n^2 - 1$ and $(n^2 - 1) + 5$ are divisible by 5.

This implies that 5^2 divides $5(5^{a-1} - 1)$, i.e. 5 divides $5^{a-1} - 1$. This can hold if and only if $5^{a-1} - 1 = 0$, that is if and only if $a = 1$, and hence $n = 1$.

From (1)–(3) it follows (see Problems 36 and 38) that a_n has an infinite decimal expansion for all natural numbers n except $n = 1$; if $n \neq 5k \pm 1$ for $k = 1, 2, \ldots$, then the decimal expansion is purely periodic.

Problem 45

For any natural number m

$$\frac{k_1}{10} + \frac{k_2}{10^2} + \ldots + \frac{k_m}{10^m} \leqslant \frac{r}{s} < \frac{k_1}{10} + \frac{k_2}{10^2} + \ldots + \frac{k_m}{10^m} + \frac{1}{10^m},$$

that is

$$0 \leqslant \frac{r}{s} - \left(\frac{k_1}{10} + \frac{k_2}{10^2} + \ldots + \frac{k_m}{10^m} \right) < \frac{1}{10^m}.$$

This implies that

$$0 \leqslant \frac{10^m r - s(10^{m-1}k_1 + 10^{m-2}k_2 + \ldots + k_m)}{s} = \sigma_m < 1.$$

Since $\sigma_m < 1$, the numerator of the fraction σ_m is less than s; therefore it is one of the numbers $0, 1, 2, \ldots, s - 1$. This holds for any $m = 1, 2, \ldots$. Hence $\sigma_1, \sigma_2, \sigma_3, \ldots$ can take on only s distinct values. Already among the first $s + 1$ numbers $\sigma_1, \sigma_2, \ldots, \sigma_s, \sigma_{s+1}$ there are at least two, equal to one another.

Problem 46

(a) Let m/n be a reduced fraction. Suppose that the corresponding lattice point $A(n, m)$ is not visible. In that case the interior of the straight-line segment OA contains a lattice point $B(\ell, k)$; here O denotes the origin of the coordinate system xOy (see Fig. 2.21).

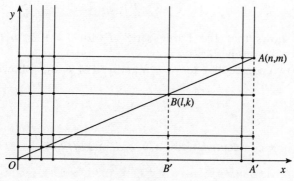

Fig. 2.21

From A and B drop perpendiculars AA' and BB' onto the x-axis. The triangles OAA' and OBB' are similar. Hence their corresponding sides are proportional:

$$\frac{m}{n} = \frac{k}{\ell}.$$

Since $k < m$ and $\ell < n$, this implies that the fraction k/ℓ is obtained by simplifying the fraction m/n. According to our assumption, m/n is a reduced fraction. This contradiction implies that the straight-line segment OA cannot contain any lattice point different from O and A. Thus A is a visible lattice point.

(b) Let $A(n, m)$ be a visible lattice point. Suppose that m/n is not a reduced fraction. In that case $m = kt$ and $n = bt$ for some integer $t > 1$. Thus

$$\frac{m}{n} = \frac{k}{\ell}, \text{ where } k < m \text{ and } \ell < n.$$

This implies that the point B with coordinates (ℓ, k) is a lattice point inside the straight-line segment OA, contradicting the assumption that A is visible. Hence m/n is a reduced fraction.

Problem 47

F_1						$\frac{0}{1}$	$\frac{1}{1}$						
F_2						$\frac{0}{1}$	$\frac{1}{2}$	$\frac{1}{1}$					
F_3					$\frac{0}{1}$	$\frac{1}{3}$	$\frac{1}{2}$	$\frac{2}{3}$	$\frac{1}{1}$				
F_4				$\frac{0}{1}$	$\frac{1}{4}$	$\frac{1}{3}$	$\frac{1}{2}$	$\frac{2}{3}$	$\frac{3}{4}$	$\frac{1}{1}$			
F_5			$\frac{0}{1}$	$\frac{1}{5}$	$\frac{1}{4}$	$\frac{1}{3}$	$\frac{2}{5}$	$\frac{1}{2}$	$\frac{3}{5}$	$\frac{2}{3}$	$\frac{3}{4}$	$\frac{4}{5}$	$\frac{1}{1}$
F_6		$\frac{0}{1}$	$\frac{1}{6}$	$\frac{1}{5}$	$\frac{1}{4}$	$\frac{1}{3}$	$\frac{2}{5}$	$\frac{1}{2}$	$\frac{3}{5}$	$\frac{2}{3}$	$\frac{3}{4}$	$\frac{4}{5}$	$\frac{5}{6}$	$\frac{1}{1}$

Inspection shows that the Farey series F_i for $i = 1, 2, \ldots, 6$ have the following striking properties.

For any two consecutive terms a/b and c/d of F_i the following relation holds:

$$bc - ad = 1. \tag{2.31}$$

and,

for any three consecutive terms a/b, c/d, and e/f of F_i

$$\frac{c}{d} = \frac{a+e}{b+f}. \tag{2.32}$$

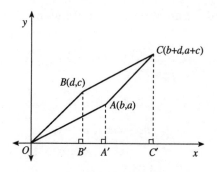

Fig. 2.22

The above properties hold for all Farey series F_i, $i = 1, 2, 3, \ldots$ as shown below.

(1) Equation (2.31) can be proved as follows. In an xOy coordinate system consider the lattice points $A(b, a)$ and $B(d, c)$ corresponding to two consecutive terms a/b and c/d of a Farey series F_i (Fig. 2.22). The construction of F_i, described on p. 45, implies that the only lattice points inside, or on the perimeter, of the triangle OAB are the three points O, A, B on its perimeter. Complete triangle OAB to a parallelogram $OACB$. The vertex C of this parallelogram is the lattice point $(b + d, a + c)$.

$ABCO$ has a centre of symmetry (the intersection of its diagonals OC and AB); all lattice points of triangle ACB are images of lattice points of OAB under this central symmetry. Hence the only lattice points of ACB are $A, C,$ and B. This implies, in view of property $\mathbb{P}$, quoted on p. 36, that

$$\text{Area } OACB = 1. \qquad (2.33)$$

On the other hand, the area of $OACB$ can be evaluated in terms of the coordinates of O, A, C, and B. For this, drop the perpendiculars AA', BB', and CC' onto the x-axis. From Fig. 2.22 we see that

$$\text{Area } OABC = \text{Area } OB'B + \text{Area } B'C'CB - \text{Area } OA'A - \text{Area } A'C'CA$$

$$= \tfrac{1}{2}cd + \tfrac{1}{2}(c + a + c)b - \tfrac{1}{2}ab - \tfrac{1}{2}(a + a + c)d$$

$$= bc - ad. \qquad (2.34)$$

(2.33) and (2.34) imply that

$$bc - ad = 1.$$

We have proved (2.31) for all Farey sequences F_i, assuming the validity of $\mathbb{P}$.

(2) If a/b, c/d, and e/f are three consecutive terms of F_i, then, in view of (2.31),

$$bc - ad = 1$$

and

$$de - cf = 1$$

Thus $$bc - ad = de - cf.$$

hence

$$c(b + f) = d(a + e),$$

that is

$$\frac{e}{d} = \frac{a + e}{b + f}.$$

(2.32) has been deduced from eqn (2.31) for all F_i, $i = 1, 2, \ldots$.

Chapter IV

Problem 48

Suppose that a light ray emerging from P meets m_1 at R, then m_2 at S, and after that passes through Q. In order to find R and S we analyse the sketch drawn in Fig. 2.23(a).

In view of the law of reflection of light,

$$\angle M_1 RP = \angle N_1 RS \quad \text{and} \quad \angle RSM_2 = \angle QSN_2 \qquad (2.35)$$

Denote by P' the reflection of P in m_1, and by Q' the reflection of Q in m_2. Clearly,

$$\angle M_1 RP = \angle M_1 RP' \quad \text{and} \quad \angle QSN_2 = \angle Q'SN_2 . \qquad (2.36)$$

From (2.35) and (2.36) it follows that

$$\angle N_1 RS = \angle M_1 RP' \quad \text{and} \quad \angle RSM_2 = \angle Q'SN_2 .$$

Thus, the points P', R, S, and Q' must be collinear.

The above analysis leads to the following construction of R and S. Reflect P in m_1 to obtain P' and reflect Q in m_2 to obtain Q'. Join P' and Q' by the straight-line segment $P'Q'$, and mark the intersections of $P'Q'$

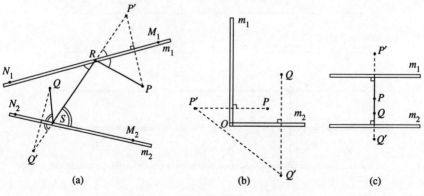

(a)   (b)   (c)

Fig. 2.23

175

with m_1 and m_2 by R and S respectively. The required path of the light ray is the polygonal line $PRSQ$.

The problem has no solution if (1) m_1 and m_2 form an angle $\alpha \neq 0°$ and $P'Q'$ does not meet the legs of this angle in two distinct points (Fig. 2.23(b)), or (2) if m_1 and m_2 are parallel and $P'Q'$ is perpendicular to m_1 (and to m_2) (see Fig. 2.23(c)). Otherwise the problem has a unique solution.

Problem 49

Figure 2.24 shows a method for the construction of the path of the light ray which emerges from $P = P_1$ and, after a total of four reflections from the mirrors m_2 and m_1, passes through $Q = Q_1$. The same method can be used for the construction of a light ray which emerges from P and, after a total of n reflections (where $n = 1, 2, \ldots$) from m_2 and m_1, passes through Q. The construction consists of the following steps.

Step 1. Draw the straight lines $m_3, m_4, \ldots, m_{n+1}$ such that m_{i+1} is the reflection of m_{i-1} in m_i for $i = 2, 3, \ldots, n$. Denote by σ_i the reflection in m_i for $i = 2, \ldots, n+1$.

Step 2. Construct the image Q_{i+1} of Q_i under σ_{i+1} for $i = 1, \ldots, n$ and join Q_{n+1} to P_1. Denote the intersection of the straight-line segment $Q_{n+1}P_1$ with m_i by P_i for $i = 2, \ldots, n+1$.

Step 3. Construct successively $P_2P'_3 = \sigma_2(P_2P_3)$, $P'_3P'_4 = \sigma_2\sigma_3(P_3P_4), \ldots,$ $P'_iP'_{i+1} = \sigma_2\sigma_3 \ldots \sigma_i(P_iP_{i+1})$, and $P'_{n+1}Q = \sigma_2 \ldots \sigma_{n+1}(P_{n+1}Q_{n+1})$. The polygonal line $P_1P_2P'_3P'_4 \ldots P'_{n+1}Q$ is the path of the required light ray.

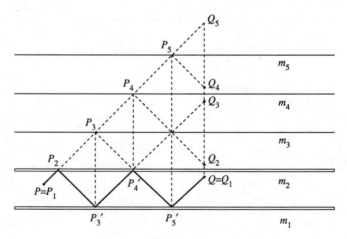

Fig. 2.24

The problem has no solution if PQ is perpendicular to the lines m_1 and m_2, and a unique solution in all other cases.

Problem 50

(a) Suppose that a light ray approaching the angle $\angle AOB$ along a given direction d meets the mirror $m_1 = OA$ at R, then the mirror $m_2 = OB$ at S, and leaves $\angle AOB$ along the direction $-d$ (Fig. 2.25(a)). This implies that

> the angle x between d and AO is equal to $\angle ORS$, and the angle y between $-d$ and OB is equal to $\angle OSR$.

From $\triangle ORS$ it follows that

$$\alpha + x + y = 180°. \tag{2.37}$$

Since d and $-d$ are parallel, the angle β between d and SR, and the angle γ between SR and $-d$, add up to $180°$. From $\beta = 180° - 2x$ and $\gamma = 180° - 2y$ it follows that

$$(180° - 2x) + (180° - 2y) = 180°,$$

that is,

$$x + y = 90°.$$

This and (2.37) yield the condition

$$\alpha = 90°.$$

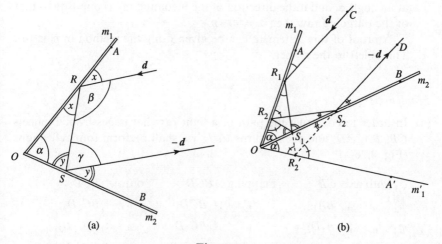

(a)

(b)

Fig. 2.25

(b) Figure 2.25(b) shows a light ray that approaches $\alpha = \angle AOB$ in a direction $\boldsymbol{d} = CR_1$, meets successively m_1 at R_1, m_2 at S_1, m_1 at R_2, m_2 at S_2, and leaves $\angle AOB$ in the direction $-\boldsymbol{d} = S_2 D$. Reflect $\angle AOB$, $R_2 S_1$, and $R_2 S_2$ in OB; their reflected images are $\angle A'OB = \alpha$, $R_2' S_1$, and $R_2' S_2$ respectively. We have

$$\angle OS_1 R_2 = \angle OS_1 R_2' \quad \text{and} \quad \angle OS_2 R_2 = \angle OS_2 R_2'. \qquad (2.38)$$

The properties of a light ray's reflection from m_2 imply that

$$\angle OS_1 R_2 = \angle R_1 S_1 S_2 \quad \text{and} \quad \angle OS_2 R_2 = \angle DS_2 B. \qquad (2.39)$$

From (2.38) and (2.39) it follows that

$$\angle R_1 S_1 S_2 = \angle OS_1 R_2' \quad \text{and} \quad \angle DS_2 B = \angle OS_2 R_2';$$

hence R_1, S_1, R_2' are collinear, and so are R_2', S_2, and D. Thus the polygonal line $\Pi = CR_1 S_1 R_2' S_2 D = CR_1 R_2' D$. Π makes equal angles with m_1 at R_1 and with m_1' at R_2'. Moreover, Π approaches $\angle AOA'$ along $CR_1 = \boldsymbol{d}$ and leaves it along $R_2' D = -\boldsymbol{d}$. In view of part (a) of this problem,

$$\angle AOA' = 90°.$$

Thus

$$\angle AOB = \tfrac{1}{2} \angle AOA' = 45°.$$

(c) The generalization of the problem in Part (a) states that:
   If a light ray is reflected n times from each of two mirrors which meet at an angle α, and if the direction of the incoming ray is opposite to that of the outgoing ray, then $\alpha = 90°/n$.
   A proof of this statement can be given using the method in part (b). This is left to the reader.

Problem 51

(a) In order to construct the path of a light ray that issues from L, meets CB, BA, AD, and then returns to L, we shall perform four reflections (Fig. 2.26(a)):

φ_1 with axis CB	mapping $ABCD$	onto $A_1 BCD_1$;
φ_2 " " BA_1	" $A_1 BCD_1$	" $A_1 BC_2 D_2$;
φ_3 " " $A_1 D_2$	" $A_1 BC_2 D_2$	" $A_1 B_3 C_3 D_2$;
φ_4 " " $D_2 C_3$	" $A_1 B_3 C_3 D_2$	" $A_4 B_4 C_3 D_2$.

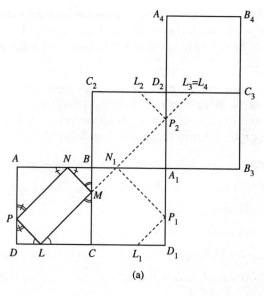

(a)

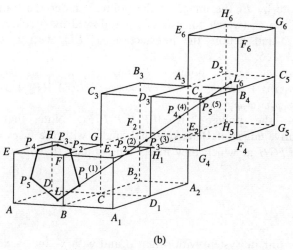

(b)

Fig. 2.26

The image of L under the transformation $\varphi_4\varphi_3\varphi_2\varphi_1$ (φ_1 first!) is L_4. Draw the straight-line segment L_4L and mark its intersections with A_1D_2, BA_1, and CB by P_2, N_1, and M respectively. Note that $(\varphi_4\varphi_3\varphi_2\varphi_1)^{-1}$ maps L_4 onto L, $(\varphi_3\varphi_2\varphi_1)^{-1}$ maps P_2 onto P on edge AD, $(\varphi_2\varphi_1)^{-1}$ maps N_1 onto N on AB, and φ_1^{-1} maps M onto itself. Join L to M, M to N, N to P, and P to L. The quadrilateral $LMNP$

has the property that any two of its consecutive sides make equal angles
with the edge on which they meet (see Fig. 2.26(a)). Thus $LMNP$ is the
path of a light ray that issues from L, meets all sides of $ABCD$, and
returns to L.

(b) The construction of a light ray that issues from a point L on $ABCD$,
meets the faces $BCFG$, $DCGH$, $FGHE$, $ADHE$, $ABFE$, and returns to
L, is a generalization of the construction performed in (a). We shall
perform six reflections in six planes (Fig. 2.26(b)):

φ_1 reflects $ABCDEFGH$ in $\pi_1 = BCGF$ and yields the image $A_1BCD_1E_1FGH_1$;

φ_2 " $A_1BCD_1E_1FGH_1$ " $\pi_2 = D_1CGH_1$ "   "   "   " $A_2B_2CD_1E_2F_2GH_1$;

φ_3 " $A_2B_2CD_1E_2F_2GH_1$ " $\pi_3 = F_2GH_4E_2$ "   "   "   " $A_3B_3C_3D_3E_2F_2GH_1$;

φ_4 " $A_3B_3C_3D_3E_2F_2GH_1$ " $\pi_4 = A_3D_3H_1E_2$ "   "   "   " $A_3B_4C_4D_3E_2F_4G_4H_1$;

φ_5 " $A_3B_4C_4D_3E_2F_4G_4H_1$ " $\pi_5 = A_3B_4F_4E_2$ "   "   "   " $A_3B_4C_5D_5E_2F_4G_5H_5$;

φ_6 " $A_3B_4C_5D_5E_2F_4G_5H_5$ " $\pi_6 = A_3B_4C_5D_5$ "   "   "   " $A_3B_4C_5D_5E_6F_6G_6H_6$.

Denote by L_6 the image of the point L under the transformation
$\varphi_6\varphi_5\varphi_4\varphi_3\varphi_2\varphi_1$ (which consists of φ_1, followed by $\varphi_2, \varphi_3, \ldots, \varphi_6$). Join
L_6 to L and denote the intersection of LL_6 with π_i by $P_i^{(i)}$ for
$i = 1, 2, \ldots, 5$.
Construct the points $P_1 = \varphi^{-1}(P_1^{(1)})$, $P_2 = (\varphi_2\varphi_1)^{-1}(P_2^{(2)})$,
$P_3 = (\varphi_3\varphi_2\varphi_1)^{-1}(P_3^{(3)})$, $P_4 = (\varphi_4\varphi_3\varphi_2\varphi_1)^{-1}(P_4^{(4)})$,   and
$P_5 = (\varphi_5\varphi_4\varphi_3\varphi_2\varphi_1)^{-1}(P_5^{(5)})$.
Draw the polygon $\Pi = LP_1P_2P_3P_4P_5$. Note that any two
consecutive edges of Π make equal angles with the faces of the cube
$ABCDEFGH$ at which they meet. This implies that Π is the required
path.

Problem 52

Set up a coordinate system with origin A and with x- and y-axes along the
sides AB and AD of the square $S = ABCD$ respectively. Take the side
length of S as 1. Then L has coordinates $(a, 0)$, where $0 < a < 1$. Divide
the plane into squares by drawing straight lines with equations
$x = k, y = k'$, for $k, k' = 0, \pm 1, \pm 2, \ldots$ (Fig. 2.27). Denote the set of these
squares by ζ.

As indicated in the hint (p. 39), the path of the light ray r inside S can be
mapped onto a straight line ℓ that passes through a sequence of adjacent
squares of ζ obtained by successive reflections of S in the lines $x = k$ and
$y = k'$.

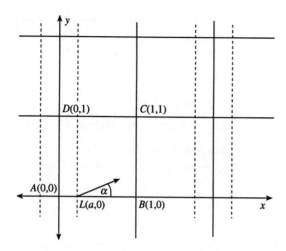

Fig. 2.27

ℓ passes through L and its gradient is $\mu = \tan \alpha$. Hence the equation of ℓ in our coordinate system is

$$y = \mu(x - a) \tag{2.40}$$

The light ray r inside S returns to L if and only if ℓ passes through a point $L_1(a + 2k, 2k')$ for some natural numbers k, k'. This will happen if and only if the coordinates of L_1 satisfy equation (2.40), that is if and only if

$$2k' = \mu \cdot 2k$$

or, equivalently, if and only if

$$\mu = \frac{k'}{k}.$$

This proves that:

r returns to L if and only if $\tan \alpha$ is a rational number, different from 0.

Denote by L_i the points with coordinates $(a + 2ki, 2k'i)$ for $i = 1, 2, \ldots$. If ℓ contains L_1, then ℓ contains all points L_i; moreover, the line segments $LL_1, L_1L_2, L_2L_3, \ldots$ all represent the image of the same portion of the path of the light ray inside the square. Thus:

If r returns to L then the path of the light ray is periodic.

Problem 53

Let $\vartheta = 1/b$, where b is a natural number. In this case, if m and n are arbitrary integers then $m\vartheta + n$ is of the form i/b for $i = 0, \pm 1, \pm 2, \ldots$. Consider the real number $r = 1/2b$. This number has a neighbourhood, for example the interval $(1/4b, 3/4b)$, which contains no element of the form i/b, indeed

$$\frac{1}{4b} < \frac{i}{b} < \frac{3}{4b}$$

would imply the absurd conclusion that

$$1 < 4i < 3 \quad \text{for some} \quad i = 0, \pm 1, \pm 2, \dots .$$

Thus Kronecker's theorem is not true for $\vartheta = 1/b$.

By similar arguments one can prove that the theorem is not true for any rational number a/b.

Problem 54

The reflected light ray will pass arbitrarily near to any point $P(\xi, \eta)$ of S if and only if ℓ passes arbitrarily near to an image P' of P with coordinates of the form $(\xi + 2m, \eta + 2n)$ for some natural numbers m and n.

The line ℓ has parametric equations

$$x = a + \lambda t, \ y = vt, \quad \text{where} \quad v/\lambda = \tan \alpha.$$

Our aim is to find a point $Q'(a + \lambda t, vt)$ on ℓ and an image $P'(\xi + 2m, \eta + 2n)$ of P that are arbitrarily near to one another. In other words, we are looking for Q' and P' whose corresponding coordinates differ by less than any small positive ε:

$$|a + \lambda t - (\xi + 2m)| < \varepsilon \tag{2.41}$$

and

$$|vt - (\eta + 2n)| < \varepsilon. \tag{2.42}$$

The equation $vt - (\eta + 2n) = 0$ yields

$$t = \frac{\eta + 2n}{v}$$

which satisfies inequality (2.42). Substitution of this value of t in (2.41) yields

$$\left| \underbrace{\frac{\lambda}{v} n}_{} - \underbrace{\left(-\eta \frac{\lambda}{v} - \frac{1}{2} a + \frac{1}{2} \xi \right)}_{} - m \right| < \frac{\varepsilon}{2},$$

$$| \ \vartheta n \ - \qquad\qquad r \qquad\qquad - m | < \tfrac{\varepsilon}{2}.$$

We have to find integers n and m satisfying the last inequality. According to our assumption, $\vartheta = 1/\tan \alpha$ is irrational. In view of Kronecker's theorem, there exist integers m and n, where n is greater than a positive number N, such that

$$r - \frac{\varepsilon}{2} < \vartheta n - m < r + \frac{\varepsilon}{2}$$

for any real number r. The above arguments imply that for these values of n and m the points $P'(\xi + 2m, \eta + 2n)$ and $Q'(a + (\lambda/v)(n + 2n), \eta + 2n)$ are arbitrarily near to one another. This proves that the path of the light ray in S passes through the point $Q'(a + (\lambda/v)(n + 2n) - 2m, n)$ which is arbitrarily near to $P(\xi, \eta)$.

Problem 55

We are looking for a natural number k such that the number 11^k starts with the digits 15101051. This means that for a certain natural number n

$$15101051 \cdot 10^n < 11^k < 15101052 \cdot 10^n. \tag{2.43}$$

By taking logarithms to the base 10, the inequalities (2.43) are transformed into

$$\log_{10} 15101051 + n \log_{10} 10 < k \log_{10} 11 < \log_{10} 15101052 + n \log_{10} 10,$$

that is, into

$$\log_{10} 15101051 < k \log_{10} 11 - n < \log_{10} 15101052. \tag{2.44}$$

Put $\vartheta = \log_{10} 11$ and $r = \frac{1}{2}(\log_{10} 15101051 + \log_{10} 15101052)$. Since ϑ is irrational, Kronecker's theorem implies the existence of integers k and n, $k > N > 0$, such that

$$r - \frac{\varepsilon}{2} < k\vartheta - n < r + \frac{\varepsilon}{2}$$

for a small positive number ε. Since $k\vartheta > 0$, this implies that $n > 0$. Clearly these values of k and n satisfy (2.44).

We have proved that there exist natural numbers k and n satisfying the original set of inequalities (2.43). The power 11^k starts with the digits 15101051.

Problem 56

Let A be the number with digits $a_1, a_2, \ldots, a_s$ in this order. We are looking for a square number n^2 whose first digits are $a_1 a_2 \ldots a_s$. This means that we want to find out whether there are natural numbers n and k satisfying the inequalities

$$A \cdot 10^k < n^2 < (A + 1)10^k, \tag{2.45}$$

where $A = a_s + a_{s-1} \cdot 10 + a_{s-2} \cdot 10^2 + \ldots a_2 \cdot 10^{s-2} + a_1 \cdot 10^{s-1}$.

The inequalities (2.45) are equivalent to

$$\log_{10} A < 2 \log_{10} n - k < \log_{10}(A + 1). \tag{2.46}$$

In order to be able to apply Kronecker's theorem, we shall consider the special case when $n = 2^m$ and $k = 2\ell$ for some natural numbers m and ℓ. This transforms (2.46) into

$$\tfrac{1}{2} \log_{10} A < m \log_{10} 2 - \ell < \tfrac{1}{2} \log_{10} (A + 1).$$

Take $r = \tfrac{1}{2} \left[\tfrac{1}{2} \log_{10} A + \tfrac{1}{2} \log_{10}(A + 1) \right]$ and put $\vartheta = \log_{10} 2$. The number ϑ is irrational. According to Kronecker's theorem, there are natural numbers m and ℓ for which

$$r - \frac{\varepsilon}{2} < m\vartheta - \ell < r + \frac{\varepsilon}{2},$$

where ε is a small, positive number. It follows that these values of m and ℓ satisfy the inequalities (2.45):

$$A \cdot 10^{2\ell} < (2^m)^2 < (A + 1) \cdot 10^{2\ell}.$$

Thus $(2^m)^2$ starts with the digits $a_1 a_2 \ldots a_s$.

Problem 57

(a) Suppose that there are real numbers x for which

$$f(x) = \sin \pi x + \sin \pi\sqrt{2} x = 2.$$

Since $\sin \alpha \leqslant 1$ for any α, this implies that

$$\sin \pi x = 1 \text{ and } \sin \pi\sqrt{2} x = 1$$

for *the same x*.

The first of these equations implies that for some integer n $\pi x = \pi/2 + 2n\pi$, that is $x = \tfrac{1}{2} + 2n$.

Hence x is a rational number.

From the second equation it follows that for some integer m $\pi\sqrt{2} x = \pi/2 + 2m\pi$, that is $x = (1/\sqrt{2})(\tfrac{1}{2} + 2m)$. This means that x is irrational.

Since the same number x cannot be both rational and irrational, we have reached a contradiction. Thus $f(x)$ cannot be equal to 2 for any real number x.

(b) Take $x = \tfrac{1}{2} + 2m$, where m is to be an integer, such that

$$f\left(\tfrac{1}{2} + 2m\right) = \sin \pi \left(\tfrac{1}{2} + 2m\right) + \sin \pi\sqrt{2} \left(\tfrac{1}{2} + 2m\right)$$

is very close to 2.

Since $\sin \pi\left(\frac{1}{2} + 2m\right) = 1$, this implies that $\sin \pi\sqrt{2}\left(\frac{1}{2} + 2m\right)$ must be very close to 1, that is $\sqrt{2}\left(\frac{1}{2} + 2m\right)$ must be very close to $\frac{1}{2} + 2n$ for some integer n.

We must find integers m and n such that

$$\left(\tfrac{1}{2} + 2n\right) - \varepsilon < \sqrt{2}\left(\tfrac{1}{2} + 2m\right) < \left(\tfrac{1}{2} + 2n\right) + \varepsilon \text{ for a small positive } \varepsilon,$$

i.e. such that

$$\frac{1}{2} - \frac{\sqrt{2}}{2} - \varepsilon < 2\sqrt{2}m - 2n < \frac{1}{2} - \frac{\sqrt{2}}{2} + \varepsilon,$$

or equivalently, such that

$$\frac{1 - \sqrt{2}}{4} - \frac{\varepsilon}{2} < \vartheta m - n < \frac{1 - \sqrt{2}}{4} + \frac{\varepsilon}{2}, \quad \text{where} \quad \vartheta = \sqrt{2}.$$

In view of Kronecker's theorem, there exist natural numbers m and n satisfying the last inequalities (since ϑ is an irrational number). Thus, for these values of m and n, $f\left(\frac{1}{2} + 2m\right)$ is arbitrarily close to 2.

Problem 58

Since $f(x) = f(x + 1)$ for all $x \in \mathbb{R}$, it follows that $f(x) = f(x + m)$ for all $x \in \mathbb{R}$ and all integers m.

From $f(x) = f(x + \vartheta)$ for all $x \in \mathbb{R}$, it follows that $f(x) = f(x + n\vartheta)$ for all $x \in \mathbb{R}$ and for all integers n.

Thus

$f(x) = f(x + m + n\vartheta)$ for all $x \in \Re$ and for all integers m and n.

Put $y = m + n\vartheta$. The functions $f(x)$ and $f(x + y)$ have the same values for all $y \in \{m + n\vartheta\}$, m and n integers. In view of Kronecker's theorem the set $\{m + n\vartheta\}$ is dense in $\mathbb{R}$. According to statement (*) on p. 44, it follows that

$$f(x) = f(x + y) \quad \text{for all real numbers } x \text{ and } y. \tag{2.47}$$

Take any two different real numbers x_1 and x_2 and put $x = x_1$ and $y = x_2 - x_1$. Equation (2.47) implies that

$$f(x_1) = f\left[x_1 + (x_2 - x_1)\right] = f(x_2).$$

Since $f(x_2)$ must be equal to $f(x_1)$ for any $x_2 \in \mathbb{R}$, it follows that $f(x)$ is a constant.

Thus all solutions of the equation

$$f(x) = f(x+1) = f(x+\vartheta) \quad \text{for all} \quad x \in \mathbb{R}$$

are of the form

$$f(x) = c,$$

where c is an arbitrary real constant number.

Chapter V

Problem 59

(a) (i) Suppose that $a \equiv b \pmod{m}$. This means

$$a = mt + r \quad \text{and} \quad b = mt' + r$$

For some integers t, t' and r.
Hence

$$a - b = (mt + r) - (mt' + r) = m(t - t'),$$

that is, m divides $a - b$.

(ii) If m divides $a - b$ then $a - b = ms$ for some integer s. Suppose that when b is divided by m the remainder is r, that is, $b = mt' + r$ for some integer t'. Since $a = b + ms$, this implies that

$$a = (mt' + r) + ms = m(t' + s) + r.$$

Hence a and b have the same remainder r when divided by m. But then

$$a \equiv b \pmod{m}$$

(b) (i) In view of (a), $a_i \equiv b_i \pmod{m}$ means that $a_i - b_i = ms_i$ for some integer s_i. Thus, according to our assumptions

$$a_1 - b_1 = ms_1$$
$$a_2 - b_2 = ms_2$$
$$+ \quad \ldots \ldots \ldots \ldots$$
$$a_n - b_n = ms_n$$

$$(a_1 + a_2 + \ldots + a_n) - (b_1 + b_2 + \ldots + b_n) = m(s_1 + s_2 + \ldots + s_n).$$

Since $s_1 + s_2 + \ldots + s_n$ is an integer,

$$a_1 + a_2 + \ldots + a_n \equiv (b_1 + b_2 + \ldots + b_n) \pmod{m}.$$

(ii) $a_1 a_2 \ldots a_n = (b_1 + ms_1)(b_2 + ms_2) \ldots (b_n + ms_n).$

The right-hand side of the last identity can be rewritten as $b_1 b_2 \ldots b_n + mc$, where c is an integer.

Thus

$$a_1 a_2 \ldots a_n \equiv b_1 b_2 \ldots b_n \,(\mathrm{mod}\ m).$$

Problem 60

(a) Since $a - a = 0 = m \cdot 0$, $a \equiv a\,(\mathrm{mod}\ m)$.

(b) $a \equiv b\,(\mathrm{mod}\ m)$ implies that $a - b = ms$ for some integer s; hence $b - a = m(-s)$. Therefore $b \equiv a\,(\mathrm{mod}\ m)$.

(c) $a \equiv b\,(\mathrm{mod}\ m)$ and $b \equiv c\,(\mathrm{mod}\ m)$ imply that $a - b = ms$ and $b - c = mt$ for some integers s and t. Therefore

$$a - c = (a - b) + (b - c) = ms + mt = m(s + t).$$

But then $a \equiv c\,(\mathrm{mod}\ m)$.

Problem 61

(a) When an arbitrary integer is divided by m, the remainder is one of the numbers $i \in S = \{0, 1, 2, \ldots, m - 1\}$. Suppose that when b is divided by m the remainder is $k \in S$. In order to find out whether the congruence $ax \equiv b\,(\mathrm{mod}\ m)$ has a solution in S, consider the remainders of the products $a \cdot 0$, $a \cdot 1$, $a \cdot 2, \ldots, a \cdot (m - 1)$ when divided by m. Suppose that division by m of the products $a \cdot i$ and $a \cdot i'$, i, $i' \in S$, $i \neq i'$ yields the same remainders. This would mean that $ai - ai' = mt$, that is $a(i - i') = mt$ for some integer t. Since m and a are coprime, m must divide $i - i'$. However, $|i - i'| < m$, and therefore m cannot divide $i - i'$. We have reached a contradiction.

Thus the remainders of the division of $a \cdot 0$, $a \cdot 1$, $a \cdot 2, \ldots, a \cdot (m - 1)$ by m must all be different. Hence one, and only one, of these products, say $a \cdot j$, has the same remainder k as the number b. The number j is a solution of the congruence

$$ax \equiv b\,(\mathrm{mod}\ m).$$

The remaining solutions of this congruence are the elements of the congruence class $\bar{j}$.

(b) Let d be the greatest common factor of a and m; this implies that $a = d \cdot a'$ and $m = d \cdot m'$ for some integers a' and m'. Hence the congruence $ax \equiv b\,(\mathrm{mod}\ m)$ can be rewritten in the form

$$da' \equiv b\,(\mathrm{mod}\ dm'),$$

that is

$$da'x - b = dm's \quad \text{for some integer } s.$$

The above equality implies that b must be divisible by d, contrary to our assumption. Therefore the congruence has no solution.

(c) Suppose that the greatest common factor d of a and m divides b, that is, $a = da'$, $m = dm'$, and $b = db'$ for some integers a', m', and b'.

$$ax \equiv b \pmod{m} \tag{2.48}$$

is equivalent to

$$ax - b = ms \quad \text{for some integer } s.$$

This yields

$$da'x - db' = dm's,$$

that is

$$a'x - b' = m's.$$

a' and m' are coprime. Hence, in view of (a), the last equation has a unique solution modulo m'. Denote this solution by s. Consider the set

$$D = \left\{ s, s + \frac{m}{d}, s + \frac{2m}{d}, \dots, s + \frac{(d-1)m}{d} \right\}.$$

It is easy to verify that all elements of D are different mod m, and that all elements of D satisfy (2.48). Hence (2.48) has at least d solutions mod m.

Suppose that (2.48) has solutions, not in D. If t is such a solution then $at \equiv b \pmod{m}$ and $as \equiv b \pmod{m}$ imply that

$$at \equiv as \pmod{m},$$

that is

$$t \equiv s \bmod \left(\frac{m}{d} \right).$$

Thus

$$t = s + \frac{m}{d} k.$$

But $k = r + d\ell$ for some integers ℓ and r, where $0 \leqslant r < d$. Hence

$$t = s + \frac{m}{d}(r + d\ell) = s + m\ell + \frac{m}{d} r \equiv \left(s + \frac{m}{d} r \right) \pmod{m}.$$

Since $s + (m/d)r \in D$, t belongs to one of the d congruence classes modulo m corresponding to the d elements of D (namely to $s + (m/d)r$). So (2.48) has exactly d different solutions modulo m.

Problem 62

Form the numbers $M_i = M/m_i = m_1 m_2 \ldots m_n/m_i$ for $i = 1, \ldots, n$. M_i and m_i are coprime for all $i = 1, 2, \ldots, n$. Hence (according to Problem 61) each of the congruences

$$M_i x \equiv 1 \,(\mathrm{mod}\ m_i)$$

has a unique solution modulo m_i. Call this solution M'_i. Construct the number

$$x^* = b_1 M_1 M'_1 + b_2 M_2 M'_2 + \ldots b_n M_n M'_n.$$

For each fixed $i \in \{1, \ldots, n\}$ consider the summands of x^* modulo m_i. For each $j \in \{1, \ldots, n\}, j \neq i$, the number M_j is divisible by m_i. Thus

$$b_j M_j M'_j \equiv 0 \,(\mathrm{mod}\ m_i) \quad \text{for} \quad j \neq i.$$

On the other hand,

$$M_i M'_i \equiv 1 \,(\mathrm{mod}\ m_i)$$

implies that

$$b_i M_i M'_i \equiv b_i \,(\mathrm{mod}\ m_i)$$

Thus

$$x^* \equiv b_i M_i M'_i \,(\mathrm{mod}\ m_i) \equiv b_i \bmod (m_i) \quad \text{for} \quad i = 1, 2, \ldots, n;$$

in other words x^* is a solution of our system of congruences.

It remains to show that x^* is unique modulo $m_1 m_2 \ldots m_n$. Suppose that our system has two solutions: x^* and y. Then $x^* \equiv y \,(\mathrm{mod}\ m_i)$ for all $i = 1, 2, \ldots, n$, that is, $x^* - y$ is divisible by m_i for all $i = 1, 2, \ldots, n$. Since the numbers m_i are pairwise coprime, this implies that $x^* - y$ is divisible by the product $M = m_1 m_2 \ldots m_n$. But then

$$x^* \equiv y \,(\mathrm{mod}\ M).$$

This completes the proof.

Problem 63

(a) Form the product $P = \displaystyle\prod_{\substack{i,j=1 \\ i \neq j}}^{n} |a_i - a_j|$

and consider the infinite set $\{P, 2P, 3P, 4P, \ldots\}$.
The numbers $a_1 + kP, a_2 + kP, \ldots, a_n + kP$ are pairwise coprime for all $k = 1, 2, 3, \ldots$. Suppose this is false. Then for some $i, j \in \{1, 2, \ldots, n\}$, $i \neq j$, the numbers $a_i + kP$ and $a_j + kP$ have a common divisor $d > 1$.

This implies that $a_i - a_j = (a_i + kP) - (a_j + kP)$ is divisible by d, and therefore P is divisible by d.

However, if $a_i + kP$ and P are both divisible by d, then a_i must also be divisible by d. Similarly, since $a_j + kP$ and P are both divisible by d, the number a_j is also divisible by d. Thus $d > 1$ is a common factor of a_i and a_j, contradicting the assumption that a_i and a_j are coprime.

Hence the numbers $a_i + kP$, $i = 1, 2, \ldots, n$ are pairwise coprime.

(b) Denote by D the difference of the greatest and smallest of the numbers $a_1, a_2, \ldots, a_n$. Suppose that the stated conditions are satisfied. Consider all prime numbers $p \leqslant D$. For each such p find an r_p that does not occur more than once in the set of remainders of $a_1, a_2, \ldots, a_n$ divided by p. Put

$$r_p^* = \begin{cases} p - r_p & \text{if } r_p > 0 \\ 0 & \text{if } r_p = 0. \end{cases}$$

Since distinct primes are pairwise coprime, the Chinese remainder theorem implies that there exists an integer b such that the congruences

$$b \equiv r_p^* \,(\text{mod } p)$$

are satisfied for all primes $p \leqslant D$.

Our next step is to prove that the numbers $a_1 + b, a_2 + b, \ldots, a_n + b$ are pairwise coprime.

Suppose that $a_i + b$ and $a_j + b$ have a common prime divisor $\bar{p}$ for some $i, j \in \{1, 2, \ldots, n\}$, $i \neq j$. Two cases will be distinguished.

(i) $\bar{p} > D$. In this case

$$a_i + b \equiv 0 \,(\text{mod } \bar{p})$$

and

$$a_j + b \equiv 0 \,(\text{mod } \bar{p})$$

imply that $a_i - a_j$ is divisible by $\bar{p}$, that is, $|a_i - a_j| \geqslant \bar{p}$. This contradicts $\bar{p} > D > |a_i - a_j|$.

Thus case (i) cannot occur.

(ii) Consider the case when $\bar{p}$ is a prime $p \leqslant D$. Then $a_i + b$ and $a_j + b$ can be rewritten as

$$a_i + b = [mp + r_i(p)] + [np + r_p^*], \text{ and}$$

$$a_j + b = [\ell p + r_j(p)] + [np + r_p^*],$$

where m, n, ℓ are integers and $r_i(p)$ and $r_j(p)$ are the respective remainders of the division of a_i and a_j by p.

Suppose that $r_p^* = p - r_p$. Then $a_i + b$ is divisible by p only if $r_i(p) - r_p$ is divisible by p, that is, if $r_i(p) = r_p$. Similarly, $a_j + b$ is divisible by p if $r_j(p) = r_p$. However, r_p appears at most once in the set of remainders of the division of $a_1, a_2, \ldots, a_n$ by p. Hence this cannot happen.

If $r_p^* = 0$ then $r_i(p)$ and $r_j(p)$ would both be divisible by p. This would lead to $r_i(p) = r_j(p) = 0 = r_p$ and would again contradict the choice of r_p.

(i) and (ii) imply that $a_1 + b, a_2 + b, \ldots, a_n + b$ are pairwise coprime.

Form the product

$$P = \prod_{\substack{i,j=1 \\ i \neq j}}^{n} |(a_i + b) - (a_j + b)|.$$

According to part (a) of Problem 63, the numbers $a_1 + (b + kP^*)$, $a_2 + (b + kP^*), \ldots, a_n + (b + kP^*)$ are pairwise coprime.

Problem 64

We shall use the notation introduced in the 'Hint' for the solution of the problem on p. 49.

The numbers $m_1, m_2, \ldots, m_n$ are pairwise coprime. According to the Chinese remainder theorem, system (I) has a solution $x = a$ unique modulo $m_1 m_2 \ldots m_n$. Similarly, since the numbers $M_1, M_2 \ldots M_n$ are pairwise coprime, system (II) has a solution $y = b$ unique modulo $M_1 M_2 \ldots M_n$.

Since a and b are integers, (a, b) is a lattice point. Consider all lattice points $(a + r, b + s)$, $r, s \in \{1, 2, \ldots, n\}$. To prove that each point $(a + r, b + s)$ is non-visible, we must show that its coordinates $a + r, b + s$ are not coprime.

The rth congruence of system (I) can be rewritten as

$$a + r \equiv 0 \pmod{m_r},$$

and the sth congruence of system (II) leads to

$$b + s \equiv 0 \pmod{M_s}.$$

Thus $a + r$ is divisible by m_r and $b + s$ is divisible by M_s. The number m_r is the product of the entries in the rth row of the matrix M, and M_s is the product of the entries in the sth column of M. The rth row and the sth column of M have an entry in common. This entry appears as a factor in both m_r and M_s. Thus $a + r$ and $b + s$ are not coprime. Hence the lattice point $(a + r, b + s)$ is not visible from the origin.

Chapter VI

Problem 65

Let $S = x_1 y_{i_1} + x_2 y_{i_2} + \ldots + x_j y_{i_j} + \ldots x_n y_{i_n}$.

Suppose that $y_n = y_{i_j}$. In S interchange y_{i_n} and y_{i_j}; this transforms S into $S' = x_1 y_{i_1} + x_2 y_{i_2} + \ldots + x_j y_{i_n} + \ldots x_n y_n$. The difference

$$S' - S = x_n(y_n - y_{i_n}) + x_j(y_{i_n} - y_n) = (y_n - y_{i_n})(x_n - x_j) \geqslant 0$$

because $y_n \geqslant y_{i_n}$ and $x_n \geqslant x_j$. Thus

$$S' \geqslant S.$$

Next find in S' the summand containing y_{n-1}. Suppose that $y_{n-1} = y_{i_k}$. In S' interchange y_{i_k} and $y_{i_{n-1}}$; this transforms S' into S''. It is easy to verify that

$$S'' \geqslant S'.$$

Hence $S'' \geqslant S$.

The above process can be continued until the sum $S^* = x_1 y_1 + x_2 y_2 + \ldots + x_n y_n$ is obtained, which is larger than or equal to all previous sums $S, S', S'', \ldots$.

This proves the inequality

$$x_1 y_{i_1} + x_2 y_{i_2} + \ldots x_n y_{i_n} \leqslant x_1 y_1 + x_2 y_2 + \ldots x_n y_n.$$

The inequality

$$x_1 y_{i_1} + x_2 y_{i_2} + \ldots + x_n y_{i_n} \geqslant x_1 y_n + x_2 y_{n-1} + x_n y_1$$

can be proved by a similar method. This is left to the reader.

In both relations equality holds if and only if $x_1 = x_2 = \ldots = x_n$ or $y_1 = y_2 = \ldots = y_n$.

Problem 66

(a) If $x_i' \geqslant x_j'$ then $y_i' = 1/x_i' \leqslant 1/x_j' = y_j'$. Thus in the sum $S = x_1' y_1' + x_2' y_2' + \ldots + x_n' y_n'$ the largest of the numbers $x_1', x_2', \ldots, x_n'$ is multiplied by the smallest of the numbers $y_1', y_2', \ldots, y_n'$, the second largest of the numbers $x_1', \ldots, x_n'$ is multiplied by the second smallest of $y_1', \ldots, y_n'$, and

193

so on. Hence, according to (1.24) in Problem 65 (p. 52) we have the inequality

$$x_1' y_1' + x_2' y_2' + \ldots + x_n' y_n' \leqslant x_1' y_n' + x_2' y_1' + x_3' y_2' + \ldots + x_n' y_{n-1}' \quad (2.49)$$

that is

$$1 + 1 + \ldots + 1 \leqslant \frac{a_1}{\sqrt[n]{a_1 \ldots a_n}} + \frac{a_2}{\sqrt[n]{a_1 \ldots a_n}} + \frac{a_3}{\sqrt[n]{a_1 \ldots a_n}} + \ldots$$

$$+ \frac{a_n}{\sqrt[n]{a_1 \ldots a_n}}.$$

It follows that

$$n \leqslant \frac{a_1 + a_2 + a_3 + \ldots + a_n}{\sqrt[n]{a_1 \ldots a_n}}$$

$$\sqrt[n]{a_1 a_2 \ldots a_n} \leqslant \frac{a_1 + a_2 + \ldots + a_n}{n} \quad (2.50)$$

This proves Cauchy's inequality for positive real numbers $a_1, \ldots, a_n$.

It should be pointed out that the inequality hodls for all non-negative real numbes $a_1, \ldots, a_n$. This is so because, if at least one of the a_i is 0, then the left-hand side, $\sqrt[n]{a_1 a_2 \ldots a_n}$ becomes 0, while the right-hand side of the inequality is a non-negative number.

(b) Equality will hold in (2.50) if and only if equality holds in (2.49). The latter will happen if and only if $x_1' = x_2' = \ldots = x_n'$ (see the solution of Problem 65).

$$x_i' = x_{i+1}' \quad \text{if and only if} \quad \sqrt[n]{a_1 a_2 \ldots a_n} = a_{i+1} \quad \text{for} \quad i = 1, \ldots, n-1,$$

and

$$x_n' = x_i' \quad \text{if and only if} \quad \sqrt[n]{a_1 a_2 \ldots a_n} = a_1.$$

Thus equality holds in (2.50) if and only if $a_1 = a_2 = \ldots = a_n$.

Problem 67

(a) We shall give two geometrical proofs of Cauchy's inequality for the case when $n = 2$ and a_1 and a_2 are two arbitrary, positive real numbers.

Proof 1. Let AB be a straight-line segment and D a point of AB such that AD is of length a_1 and DB of length a_2 (Fig. 2.28). Denote by O the midpoint of AB and draw a semicircle c wiith diameter AB. At D erect a perpendicular to AB, and mark its intersection with c by C.

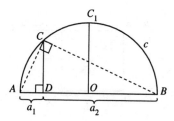

Fig. 2.28

The triangle ABC is a right-angled triangle. From the similarity of triangles ADC and CDB it follows that CD, the altitude of ABC corresponding to the hypotenuse, is equal to $\sqrt{AD \cdot DB}$. Since $CD \leqslant C_1O = AB/2$ this implies that

$$\sqrt{a_1 \cdot a_2} \leqslant \frac{a_1 + a_2}{2}.$$

Equality is reached if and only if $CD = C_1O = AO$, that is, if and only if $a_1 = a_2$.

Proof 2. This has the advantage that it can be generalized for proving Cauchy's inequality in the remaining cases, that is, when $n > 2$.

The idea for the proof is based on the following observation, connected with the case $n = 2$ and $a_1 = a_2$.

Draw a square $ABCD$ with side lengths $\sqrt{a_1} = \sqrt{a_2}$. Divide the square by its diagonal AC into two triangles ADC and ABC (Fig. 2.29(a)). Since the area of $ABCD$ is the sum of the areas of ADC and ABC, it follows that

$$\sqrt{a_1} \cdot \sqrt{a_2} = \frac{\left(\sqrt{a_1}\right)^2}{2} + \frac{\left(\sqrt{a_2}\right)^2}{2} \quad \text{if} \quad a_1 = a_2. \tag{2.51}$$

The above idea can be modified if $a_1 \neq a_2$, as shown in Fig. 2.29(b).

Suppose that $a_2 < a_1$. Draw a rectangle $ABCD$ such that AB and AD have side lengths $\sqrt{a_1}$ and $\sqrt{a_2}$ respectively. On the sides AB and DC of $ABCD$ mark the points B' and C' such that $AB'C'D$ is a square. Draw the diagonal AC' of $AB'C'D$ and denote its intersection with the straight line BC by C''. In this way two isosceles right-angled triangles are constructed: ADC' of area $\frac{1}{2}\left(\sqrt{a_2}\right)^2$ and ABC'' of area $\frac{1}{2}\left(\sqrt{a_1}\right)^2$. As Fig. 2.29(b) shows, the area $\sqrt{a_1} \cdot \sqrt{a_2}$ of the rectangle $ABCD$ is smaller than the sum of the areas of ADC' and ABC''. Thus

$$\sqrt{a_1} \cdot \sqrt{a_2} = \frac{\left(\sqrt{a_1}\right)^2}{2} + \frac{\left(\sqrt{a_2}\right)^2}{2} \quad \text{if} \quad a_1 \neq a_2. \tag{2.52}$$

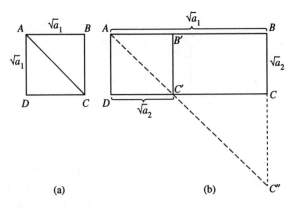

Fig. 2.29

(2.51) and (2.52) together yield Cauchy's inequality for $n = 2, a_1 > 0, a_2 > 0$:

$$\sqrt{a_1 a_2} \leqslant \frac{a_1 + a_2}{2}.$$

(b) $n = 3$. Let a_1, a_2, and a_3 be arbitrary, positive real numbers.

If $a_1 = a_2 = a_3$, draw a cube $ABCDEFGH$ of edge lengths $CB = \sqrt[3]{a_1}$, $CD = \sqrt[3]{a_2}$, and $CG = \sqrt[3]{a_3}$ (Fig. 2.30(a)). Draw the diagonals CA, CF, and CH of the cube's faces. In this way the cube is partitioned into three pyramids with square bases and common apex C: $CABFE$ of volume $\frac{1}{3}\left(\sqrt[3]{a_1}\right)^3$, $CAEHD$ of volume $\frac{1}{3}\left(\sqrt[3]{a_2}\right)^3$, and $CEFGH$ of volume $\frac{1}{3}(\sqrt[3]{a_3})^3$. The sum of these volumes is the volume $\sqrt[3]{a_1} \cdot \sqrt[3]{a_2} \cdot \sqrt[3]{a_3}$. of the cube. Thus

$$\sqrt[3]{a_1} \cdot \sqrt[3]{a_2} \cdot \sqrt[3]{a_3} = \frac{\left(\sqrt[3]{a_1}\right)^3}{3} + \frac{\left(\sqrt[3]{a_2}\right)^3}{3} + \frac{\left(\sqrt[3]{a_3}\right)^3}{3} \quad \text{if} \quad a_1 = a_2 = a_3.$$

(2.53)

Consider now the case when a_1, a_2, and a_3 are not all equal. Draw a cuboid $ABCDEFGH$ with edge lengths $CB = \sqrt[3]{a_1}$, $CD = \sqrt[3]{a_2}$ and $CG = \sqrt[3]{a_3}$. Without loss of generality we may suppose that $a_2 \leqslant a_1$ and $a_2 \leqslant a_3$. On DA and CB mark points A' and B' respectively such that $DA' = \sqrt[3]{a_2}$ and $CB' = \sqrt[3]{a_2}$. Construct the cube $A'B'CDE'F'G'H'$ and divide it (as in Fig. 2.30(a)) into three pyramids: $CA'B'F'E'$, $CA'E'H'D$, and $CE'F'G'H'$.

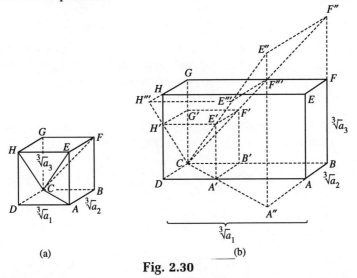

Fig. 2.30

Produce the edges CA', CF', and CE' of $CA'B'F'E'$ and denote their intersections with the plane $ABFE$ by A'', F'', and E'' respectively. Produce the edges CE', CF', and CH' and denote their intersections with the plane $EFGH$ by E''', F''', and H''' respectively (Fig. 2.30(b)).

It is easy to see that in this case

Volume $ABCDEFGH <$ Volume $CA''BF''E''$ + Volume $CA'E'H'D$

$$+ \text{ Volume } CE'''F'''GH''',$$

that is

$$\sqrt[3]{a_1} \cdot \sqrt[3]{a_2} \cdot \sqrt[3]{a_3} < \frac{\left(\sqrt[3]{a_1}\right)^3}{3} + \frac{\left(\sqrt[3]{a_2}\right)^3}{3} + \frac{\left(\sqrt[3]{a_3}\right)^3}{3}. \qquad (2.54)$$

(2.53) and (2.54) yield Cauchy's inequality for $n = 3$, $a_1 a_2 a_3 \neq 0$:

$$\sqrt[3]{a_1 a_2 a_3} \leqslant \frac{a_1 + a_2 + a_3}{3}.$$

Remark. The above proof can be generalized for $n \geqslant 4$ by comparing volumes of higher-dimensional shapes.

Problem 68

(a) (i) Suppose that $A_i = 0$ for some $i \in \{1,\ldots,k\}$. This implies that $a_{1i} = a_{2i} = \ldots = a_{ni} = 0$ and hence that $G_1 = G_2 = \ldots = G_n = 0$. But then $\sqrt[k]{A_1 A_2 \ldots A_k} = 0 = \dfrac{G_1 + G_2 + \ldots + G_n}{n}$

(ii) If $A_i > 0$ for all $1,\ldots,k$, then Cauchy's inequality applied to the numbers $a_{i1}/A_1, a_{i2}/A_2, \ldots, a_{ik}/A_k$ yields

$$\sqrt[k]{\frac{a_{i1}}{A_1} \cdot \frac{a_{i2}}{A_2} \cdot \ldots \cdot \frac{a_{ik}}{A_k}} \leqslant \frac{a_{i1}/A_1 + a_{i2}/A_2 + \ldots + a_{ik}/A_k}{k}$$

that is

$$\frac{kG_i}{\sqrt[k]{A_1 A_2 \ldots A_k}} \leqslant \frac{a_{i1}}{A_1} + \frac{a_{i2}}{A_2} + \ldots + \frac{a_{ik}}{A_k}. \tag{2.55}$$

By adding inequalities (2.55) for $i = 1, 2, \ldots, n$ we see that

$$\frac{k(G_1 + G_2 + \ldots + G_n)}{\sqrt[k]{A_1 A_2 \ldots A_k}} \leqslant \frac{nA_1}{A_1} + \frac{nA_2}{A_2} + \ldots + \frac{nA_k}{A_k} = kn,$$

that is

$$\sqrt[k]{A_1 A_2 \ldots A_k} \geqslant \frac{G_1 + G_2 + \ldots + G_n}{n}.$$

(b) In the above relations equality holds if at least one of the columns of matrix M consists entirely of zeros, or if the numbers in all rows of M are proportional.

Problem 69

The matrix M in this problem leads to $G_1 = G_2 = \ldots = G_n = \sqrt[n]{a_1 a_2 \ldots a_n}$ and $A_1 = A_2 = \ldots = A_n = (a_1 + a_2 + \ldots + a_n)/n$.

According to (1.25)

$$\sqrt[n]{\left(\frac{a_1 + a_2 + \ldots + a_n}{n}\right)^n} \geqslant \frac{n \sqrt[n]{a_1 a_2 \ldots a_n}}{n},$$

that is

$$\frac{a_1 + a_2 + \ldots + a_n}{n} \geqslant \sqrt[n]{a_1 a_2 \ldots a_n}.$$

Problem 70

Construct the matrix

$$M = \begin{pmatrix} a_1^2 & b_1^2 \\ a_2^2 & b_2^2 \\ \cdot & \cdot \\ \cdot & \cdot \\ \cdot & \cdot \\ \cdot & \cdot \\ \cdot & \cdot \\ \cdot & \cdot \\ a_n^2 & b_n^2 \end{pmatrix}.$$

According to (1.25)

$$\sqrt{\frac{a_1^2 + a_2^2 + \ldots + a_n^2}{n} \cdot \frac{b_1^2 + b_2^2 + \ldots + b_n^2}{n}} \geqslant \frac{|a_1 b_1| + |a_2 b_2| + \ldots + |a_n b_n|}{n}.$$

This leads to

$$\left(a_1^2 + a_2^2 + \ldots + a_n^2\right)\left(b_1^2 + b_2^2 + \ldots + b_n^2\right) \geqslant \left(a_1 b_1 + a_2 b_2 + \ldots a_n b_n\right)^2$$

Problem 71

The solution of this problem using (1.25) is outlined in the hint on p. 54. We leave it to the reader to complete the proof.

Problem 72

The solution, using (1.25) is outlined in the hint (p. 55).

Problem 73

(a) Consider the matrix M, described in the hint on p. 55.
   In this case

$$G_1 = G_2 = \ldots = G_{n+1} = \sqrt[n+1]{\left(1 + \frac{x}{n}\right)^n} \text{ and}$$

$$A_1 = A_2 = \ldots = A_{n+1} = \frac{n + 1 + x}{n + 1}.$$

Thus, according to (1.25)

$$\sqrt[n+1]{\left(1+\frac{x}{n+1}\right)^{n+1}} \geqslant \frac{(n+1)\sqrt[n+1]{(1+x/n)^n}}{n+1},$$

that is,

$$1+\frac{x}{n+1} \geqslant \sqrt[n+1]{\left(1+\frac{x}{n}\right)^n},$$

$$\left(1+\frac{x}{n+1}\right)^{n+1} \geqslant \left(1+\frac{x}{n}\right)^n. \tag{2.56}$$

(b) Put $x = 1$ in (2.56) and obtain

$$\left(1+\frac{1}{n+1}\right)^{n+1} \geqslant \left(1+\frac{1}{n}\right)^n.$$

Equality can be ruled out (e.g. by applying the binomial formula). This proves that

$$\left(1+\frac{1}{n+1}\right)^{n+1} > \left(1+\frac{1}{n}\right)^n \quad \text{for} \quad n = 1, 2, \ldots.$$

If $x = -1$ is put in (2.56), this becomes

$$\left(1-\frac{1}{n+1}\right)^{n+1} \geqslant \left(1-\frac{1}{n}\right)^n,$$

that is

$$\left(\frac{n}{n+1}\right)^{n+1} \geqslant \left(\frac{n-1}{n}\right)^n.$$

In the last inequality replace n by $n+1$. This gives

$$\left(\frac{n+1}{n+2}\right)^{n+2} \geqslant \left(\frac{n}{n+1}\right)^{n+1},$$

that is

$$\left(1+\frac{1}{n+1}\right)^{n+2} \leqslant \left(1+\frac{1}{n}\right)^{n+1}.$$

Again, it is not difficult to rule out equality; thus

$$\left(1+\frac{1}{n+1}\right)^{n+2} < \left(1+\frac{1}{n}\right)^{n+1} \quad \text{for} \quad n = 1, 2, \ldots .$$

Problem 74

Consider the set $\mathscr{S}$ of all rectangles with a given perimeter P. Denote the side lengths of an arbitrary rectangle $R \in \mathscr{S}$ by x and y; the side length of the square $S \in \mathscr{S}$ is $P/4 = (x+y)/2$.

The area of R is xy and the area of S is $((x+y)/2)^2$. According to Cauchy's inequality applied to x and y,

$$\sqrt{xy} \leqslant \frac{x+y}{2},$$

and equality holds if and only if $x = y$.

This implies that

$$xy \leqslant \left(\frac{x+y}{2}\right)^2 = \left(\frac{P}{4}\right)^2,$$

that is, of all rectangles with a given perimeter the square has the largest area.

Problem 75

Let $\mathscr{T}$ be the set of all triangles with a given perimeter $2s$. Denote the side lengths of an arbitrary triangle $T \in \mathscr{T}$ by a, b, and c. The side length of the equilateral triangle $E \in \mathscr{T}$ is $2s/3$.

By applying Cauchy's inequality to the numbers $s - a$, $s - b$ and $s - c$ we find that

$$\sqrt[3]{(s-a)(s-b)(s-c)} \leqslant \frac{(s-a)+(s-b)+(s-c)}{3}, \tag{2.57}$$

that is

$$(s-a)(s-b)(s-c) \leqslant \left(\frac{s}{3}\right)^3.$$

The latter inequality can be transformed into

$$s(s-a)(s-b)(s-c) \leqslant \frac{s^4}{27}, \tag{2.58}$$

or

$$\sqrt{s(s-a)(s-b)(s-c)} \le \left(\frac{2s}{3}\right)^2 \frac{\sqrt{3}}{4}.$$

Equality in (2.57) and hence in (2.58) is achieved if and only if $s-a = s-b = s-c$, that is, if and only if $a = b = c$.

Since $\sqrt{s(s-a)(s-b)(s-c)}$ is the formula for the area of a triangle with sides a,b,c and perimeter $2s$, and since $(2s/3)^2 \sqrt{3}/4$ is the area of an equilateral triangle of perimeter $2s$, the above relation implies that among all triangles with perimeter $2s$, the equilateral triangle has the greatest area.

Problem 76

First we shall prove the following statement.

(S) Among all triangles sharing the same base, and having the same perimeter, the isosceles triangle has the greatest area.

Proof of (S). Let $\mathcal{T}$ be the set of all triangles with given base a, and given perimeter $2s$. Denote the two remaining sides of an arbitrary triangle $T \in \mathcal{T}$ by b and c.

The area of T is $\sqrt{s(s-a)(s-b)(s-c)}$. In this expression $s(s-a)$ is constant. Thus T will have the largest area if and only if $\sqrt{(s-b)(s-c)}$ attains its maximum. According to Cauchy's inequality applied to $s-b$ and $s-c$,

$$\sqrt{(s-b)(s-c)} \le \frac{2s-b-c}{2} = \frac{a}{2},$$

and equality holds if and only if $b = c$. In other words, the area of T will be largest when T is isosceles.

This proves (S).

Consider now all quadrilaterals with a given perimeter P. Clearly, a concave quadrilateral (that is, a quadrilateral with a reflex angle) will have a smaller area than a convex quadrilateral with the same perimeter (Fig. 2.31(a)). Hence we shall restrict our attention to convex quadrilaterals of perimeter P.

Let $ABCD$ be an arbitrary convex quadrilateral of perimeter P. Our first task is to compare the area of $ABCD$ with the area of the quadrilateral $A'BC'D$ constructed as follows.

$A'BC'D$ has the same diagonal BD as $ABCD$; the triangles $A'BD$ and $C'BD$ into which BD divides $A'BC'D$ are isosceles: $A'B = A'D$ and $C'B = C'D$. Moreover, $A'BD$ has the same perimeter as ABD, and $C'BD$ has the same perimeter as CBD (Fig. 2.31(b)).

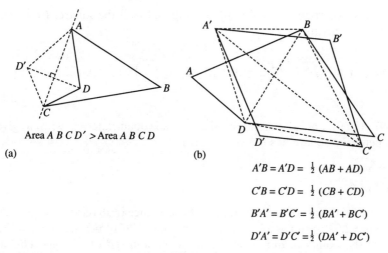

$A'B = A'D = \frac{1}{2}(AB + AD)$

$C'B = C'D = \frac{1}{2}(CB + CD)$

$B'A' = B'C' = \frac{1}{2}(BA' + BC')$

$D'A' = D'C' = \frac{1}{2}(DA' + DC')$

Fig. 2.31

In view of statement (S),

$$\text{Area } ABCD = \text{Area } ABD + \text{Area } CBD$$

$$\leqslant \text{Area } A'BD + \text{Area } C'BD$$

$$= \text{Area } A'BC'D. \tag{2.59}$$

Since $A'BC'D$ is symmetrical with respect to $A'C'$, the lengths of $A'B + BC'$ and of $A'D + DC'$ are equal to $P/2$.

The next step is to compare the area of $A'BC'D$ with the area of the rhombus $A'B'C'D'$ which has the same perimeter P and the same diagonal $A'C'$ as $A'BC'D$.

The isosceles triangle $A'B'C'$ has the same perimeter as triangle $A'BC'$ and the isosceles triangle $A'D'C'$ has the same perimeter as $A'DC'$. Hence, according to statement (S)

$$\text{Area } A'BC'D = \text{Area } A'BC' + \text{Area } A'DC'$$

$$\leqslant \text{Area } A'B'C' + \text{Area } A'D'C'$$

$$= \text{Area } A'B'C'D'. \tag{2.60}$$

Finally, the rhombus $A'B'C'D'$ is compared with the square $A'B'C''DP''$ of the same perimeter P:

$$\text{Area } A'B'C'D' = \left(\frac{P}{4}\right)^2 \sin \angle B'A'D' \leqslant \left(\frac{P}{4}\right)^2 \cdot 1 = \text{Area } A'B'C''D''.$$

(2.61)

(2.59), (2.60) and (2.61) imply that of all quadrilaterals with a given perimeter the square has the largest area.

Problem 77

The idea of the proof is illustrated in the sequence of diagrams in Fig. 2.32.

Figure 2.32(a) shows an equilateral triangle ABC of perimeter $3a$. In Fig. 2.32(b) the points E and D on the sides AC and AB of triangle ABC are such that $AE = \frac{3}{4}AC$ and $AD = \frac{3}{4}AD$. The point F is constructed so that $ADFE$ is a rhombus. The perimeter of $ADFE$ is $4 \cdot \frac{3}{4}a = 3a$ which is the same as the perimeter of ABC. However, as indicated in the diagram,

$$\text{Area } ADFE = \text{Area } ABC + \text{Area } \boxed{*} > \text{Area } ABC. \quad (2.62)$$

Figure 2.32(c) depicts a rectangle $ADGH$ having the same base and the same height as the rhombus $ADFE$; $ADIJ$ is a square with base AD.

$$\text{Area } ADIJ = \text{Area } ADGH + \text{Area } HGIJ$$

$$= \text{Area } ADFE + \text{Area } HGIJ$$

$$> \text{Area } ADFE. \quad (2.63)$$

The perimeter of the square $ADIJ$ is $4 \cdot \frac{3}{4}a = 3a$.

In view of (2.62) and (2.63), the area of a square having the same perimeter as an equilateral triangle T is greater than the area of T.

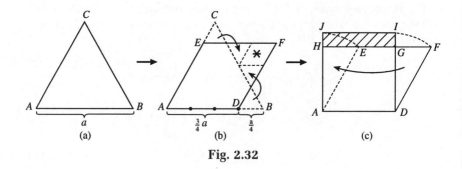

Fig. 2.32

Problem 78

Let Π' be a polygon with n equal sides of length a, and let $\Pi = A_1A_2\ldots A_n$ be the regular n-gon of side length a.

Construct the circle c passing through the vertices of Π. The sides of Π divide the interior of c into $n+1$ parts: the interior of Π and n congruent segments s (shaded in Fig. 2.33) bounded by the side A_iA_{i+1} of Π and the smaller of the arcs $\overset{\frown}{A_iA_{i+1}}$ for $i = 1,\ldots,n$ ($A_{n+1} = A_1$).

Let us construct a segment s', congruent to s, above each side of Π', outside Π' (as shown in Fig. 2.33(a)). If Π' is convex then no two segments s attached to it overlap. Denote the shape, composed of Π' and the n attached segments s', by Π^*. The perimeter of Π^* is equal to the circumference of c. Hence in view of the isoperimetric theorem,

$$\text{Area } \Pi^* \leqslant \text{Area } c;$$

that is

$$\text{Area } \Pi' + n \cdot \text{Area } s' \leqslant \text{Area } \Pi + n \cdot \text{Area } s.$$

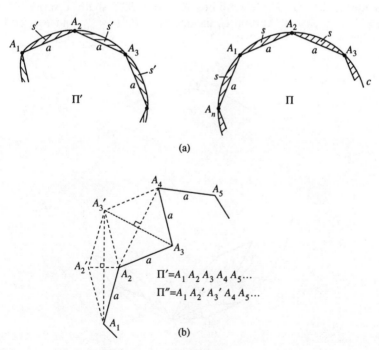

(a)

(b)

Fig. 2.33

Hence

$$\text{Area } \Pi' \leqslant \text{Area } \Pi.$$

If Π' is not convex then, by reflecting parts of it in appropriate diagonals (see Fig. 2.33(b)), one can construct a convex polygon Π'' with side lengths a whose area is larger than the area of Π'. By attaching n segments congruent to s to Π'' we can show, as above, that the area of Π'' is not larger than the area of Π. This finishes the proof of the statement:

among all n-gons with equal sides of length a the regular n-gon has the largest area.

Problem 79

(a) Draw $n-1$ non-overlapping rhombuses $R_1^{(1)}, R_2^{(1)}, \ldots, R_{n-1}^{(1)}$ of side length 1, with acute angles α, and a common vertex A_{2n}, so that $R_i^{(1)}$ and $R_{i+1}^{(1)}$ share a side $A_{2n}B_i$ for $i = 1, 2, \ldots, n-2$.

The angle of $R_i^{(1)}$ at B_i and the angle of $R_{i+1}^{(1)}$ at B_i add up to $2(180° - \alpha) = 360° - 2\alpha$, leaving a 'gap' of 2α outside the two rhombuses. In this gap place a rhombus $R_i^{(2)}$ for $i = 1, 2, \ldots, n-2$ with acute angles 2α. The rhombuses $R_i^{(2)}$ and $R_{i+1}^{(2)}$ share a vertex C_i for $i = 1, 2, \ldots, n-3$. At each C_i the angles of $R_i^{(2)}$, $R_{i+1}^{(2)}$, and $R_{i+1}^{(1)}$ add up to $2(180° - 2\alpha) + \alpha = 360° - 3\alpha$ leaving a gap of 3α; this gap will be

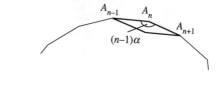

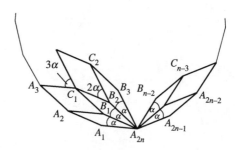

Fig. 2.34

filled by a rhombus $R_i^{(3)}$ with acute angles 3α, for $i = 1, 2, \ldots, n - 3$ (Fig. 2.34).

By continuing this process as long as there are gaps between the rhombuses of a layer a $2n$-gon $P_{2n} = A_1 A_2 \ldots A_{2n}$ is constructed.

The opposite sides of P_{2n} are parallel. This can be shown as follows. The rhombuses $R_1^{(n-1)}, R_2^{(n-2)}, \ldots, R_1^{(2)}, R_1^{(1)}$ form a 'string'. Every rhombus of this string has two sides parallel to $A_n A_{n+1}$; thus $A_{2n} A_1 \parallel A_n A_{n+1}$. The rhombuses $R_2^{(n-2)}, R_2^{(n-3)}, \ldots, R_2^{(1)}, R_1^{(1)}$ form a string in which every rhombus has two sides parallel to $A_{n+1} A_{n+2}$; thus $A_1 A_2 \parallel A_{n+1} A_{n+2}$. The rhombuses $R_3^{(n-3)}, R_3^{(n-4)}, \ldots, R_3^{(1)}, R_2^{(1)}, R_1^{(2)}$ form a string in which every rhombus has two sides parallel to $A_{n+2} A_{n+3}$; therefore $A_2 A_3 \parallel A_{n+2} A_{n+3}$. By forming similar strings of rhombuses one can prove that $A_i A_{i+1} \parallel A_{n+i} A_{n+i+1}$ for all $i = 1, 2, \ldots, n - 1$.

(b) The area of P_{2n} is the sum of the areas of the rhombuses from which it is constructed. The area of $R_i^{(j)}$ is $1 \cdot 1 \cdot \sin j\alpha$ for all $j = 1, 2, \ldots, n - 1$ and $i = 1, 2, \ldots, n - j$. Thus

$$\text{Area } P_{2n} = (n - 1) \sin \alpha + (n - 2) \sin 2\alpha + \ldots + \sin (n - 1)\alpha.$$

According to the statement proved in Problem 78, of all $2n$-gons with side length 1 the regular $2n$-gon Π_{2n} with the same side length has the largest area. The area of $\Pi_{2n} = B_1 B_2 \ldots B_{2n}$ is the sum of the areas of the $2n$ isosceles triangles with bases $B_1 B_2, B_2 B_3, \ldots, B_{2n} B_1$ and apex O, where O is the centre of Π_{2n}. Thus

$$\text{Area } \Pi_{2n} = 2n \, \frac{1 \cdot \frac{1}{2} \cot (360°/4n)}{2} = \frac{n}{2} \cot \frac{90°}{n}.$$

From Area $P_{2n} \leqslant$ Area Π_{2n} it follows that

$$(n - 1) \sin \alpha + (n - 2) \sin 2\alpha + \ldots + \sin (n - 1) \alpha \leqslant \frac{n}{2} \cot \frac{90°}{n}.$$

Hence the maximum value of $\sum_{i=1}^{n-1} (n - i) \sin i\alpha$ is $(n/2) \cot (90°/n)$.

Chapter VII

(a) Drop a perpendicular from P onto the x-axis, and denote its foot by P' (Fig. 2.35). Triangles OP_2A, OP_1A, and OPP_1' are similar (they are all right-angles triangles and share the angle $\angle P_2OA$). Hence

$$OP_2 = 2a\,\frac{OP}{OP'} = 2a\,\frac{\sqrt{x^2+y^2}}{x}\,,$$

and

$$OP_1 = 2a\,\frac{OP'}{OP} = 2a\,\frac{x}{\sqrt{x^2+y^2}}\,.$$

The above relations imply that

$$OP = OP_2 - OP_1 = 2a\,\frac{\sqrt{x^2+y^2}}{x} - 2a\,\frac{x}{\sqrt{x^2+y^2}}\,,$$

i.e.

$$\sqrt{x^2+y^2} = 2a\,\frac{\sqrt{x^2+y^2}}{x} - 2a\,\frac{x}{\sqrt{x^2+y^2}}\,,$$

This equation can be rewritten as

$$y^2 = \frac{x^3}{2a - x}\,.$$

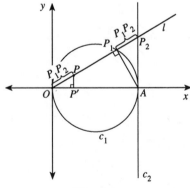

Fig. 2.35

208

(b) P^* is the intersection of the cissoid, with equation

$$y^2 = \frac{x^3}{2a - x},$$

and the straight line through $A(2a, 0)$ and $B(0, 4a)$, with equation

$$y = -2x + 4a = 2(2a - x)$$

(see Fig. 2.36).

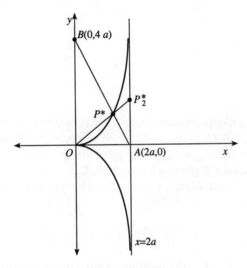

Fig. 2.36

At the intersection

$$y^2 = \frac{x^3}{2a - x} = \frac{x^3}{y/2}$$

i.e.

$$y^3 = 2x^3,$$

so

$$y = \sqrt[3]{2}x.$$

Since O, P^*, P_2^* are collinear,

$$AP_2^* = \sqrt[3]{2} \cdot OA.$$

So if $OA = 1$, then $AP_2^* = \sqrt[3]{2}$.

(c) From R draw a perpendicular RS to MN and mark the midpoint of the straight-line segment RS by O. Denote the length of RS by $2a$.

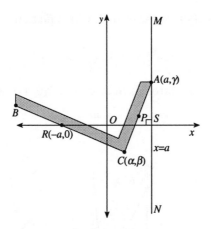

Fig. 2.37

Set up a cartesian coordinate system in which O, S, and R have the coordinates $(0,0)$, $(a,0)$, and $(-a,0)$ respectively. Figure 2.37 shows the carpenter's square ACB in an arbitrary position, with A on the line MN and with CB passing through R. Denote the coordinates of A and C by (a,γ) and (α,β) respectively. The midpoint P of AC has the coordinates

$$x = \frac{a+\alpha}{2} \quad \text{and} \quad y = \frac{\gamma+\beta}{2} \tag{2.64}$$

Since $\angle ACR = 90°$, the product of the gradients of AC and CR is -1, that is

$$\frac{\gamma-\beta}{a-\alpha} = -\frac{\alpha+a}{\beta}. \tag{2.65}$$

The length of AC is $2a$; this implies that

$$(2a)^2 = (\gamma-\beta)^2 + (a-\alpha)^2. \tag{2.66}$$

From (2.65) and (2.66) it follows that

$$\gamma^2 - \beta^2 = (a+\alpha)^2,$$

that is

$$(\gamma-\beta)(\gamma+\beta) = (a+\alpha)^2. \tag{2.67}$$

In view of (2.64), $\gamma+\beta = 2y$; therefore (2.67) can be rewritten as

$$\gamma - \beta = \frac{(a+\alpha)^2}{2y}.$$

Substitution of the above expression for $\gamma - \beta$ into (2.66) transforms the latter identity into

$$(2a)^2 = \frac{(a + \alpha)^4}{4y^2} + (a - \alpha)^2. \tag{2.68}$$

Since, according to (2.64), $\alpha = 2x - a$, eqn (2.68) yields

$$y^2 = \frac{x^3}{2a - x} \tag{2.69}$$

which is the equation of the cissoid of Diocles. Thus, as the carpenter's square ACB moves, with A gliding along MN and BC passing through R, the midpoint P of AC describes part of the cissoid with equation (2.69).

Problem 81

(a) Let $\overline{S} = ABCDEFA'B'C'D'E'F'$ be a right, regular hexagonal prism of given volume v (Fig. 2.38). Denote by O and O' the centres of the bases $ABCDEF$ and $A'B'C'D'E'F'$ respectively. On the axis OO' of $\overline{S}$ choose a point V ont he other side of O' than O at a distance $O'V < OO'$. Construct the points P on BB', Q on DD', and R on FF' such that $B'P = D'Q = F'R = O'V$. Since $A'O' = C'O' = A'B' = C'B'$, the right triangles $A'O'V$, $C'O'V$, $A'B'P$, and $C'B'P$ are congruent. Hence $VA'PC'$ is a rhombus. Similarly, $VC'QE'$ and $VE'RA$ are rhombuses.

Denote by $\tilde{S}$ the solid, bounded by the base $ABCDEF$, the six congruent trapezoids $ABPA'$, $BCC'P$, $CDQC'$, $DEE'Q$, $EFRE'$, $FAA'R$, and by the three congruent rhombuses $VA'PC'$, $VC'QE'$, and $VE'RA'$.

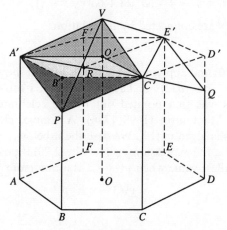

Fig. 2.38

For any position of V on OO', such that $VO' < OO'$, the volume of $\tilde{S}$ is equal to v. This is so because

$$\text{Volume } \tilde{S} = \text{Volume } \overline{S} - (\text{Volume } PA'B'C' + \text{Volume } QC'D'E'$$
$$+ \text{Volume } RE'F'A') + (\text{Volume } VA'O'C$$
$$+ \text{Volume } VC'O'E' + \text{Volume } VE'O'A),$$

where

$$\text{Volume } PA'B'C' = \text{Volume } VA'O'C'$$
$$= \text{Volume } QC'D'E' = \text{Volume } VC'O'E'$$
$$= \text{Volume } RE'F'A' = \text{Volume } VE'O'A'.$$

Our aim is to find the position of V on OO' for which the surface area of the corresponding solid $\tilde{S}$ is a minimum.

Denote the lengths of the line segments AB, $PB' = VO'$, and VP by $2a$, x, and $2y$ respectively. Denote by $\bar{s}$ the surface area of $\overline{S}$ and by $\tilde{s}$ the surface area of $\tilde{S}$. Clearly,

$$\tilde{s} = \bar{s} - (\text{Area } A'B'C'D'E'F + \text{Area } A'B'P + \text{Area } C'B'P$$
$$+ \text{Area } C'D'Q + \text{Area } E'D'Q + \text{Area } E'F'R + \text{Area } A'F'R)$$
$$+ (\text{Area } VA'PC' + \text{Area } VC'Q'E' + \text{Area } VE'R'A')$$
$$= \bar{s} - \left[6\left(\frac{2a}{2}\right)^2 \sqrt{3} + 6\frac{2ax}{2} \right] + 3\frac{2a\sqrt{3} \cdot 2y}{2}.$$

Thus

$$\tilde{s} = \bar{s} - 6a^2\sqrt{3} + 6a(y\sqrt{3} - x), \tag{2.70}$$

where x and y are connected by the relation

$$y^2 = x^2 + a^2. \tag{2.71}$$

Using (2.70) and applying calculus it is easy to show that $\tilde{s}$ reaches its minimum for $x = (a/2)\sqrt{2}$. The minimum of $\tilde{s}$ can be found without calculus; this was first pointed out by the celebrated British mathematician C. MacLaurin (1698–1746). A simple, elegant method for finding min $\tilde{s}$ is given in [19]. We describe it below.

Since $y > x$, the expression $y\sqrt{3} - x$ in (2.70) is positive. Hence $\tilde{s}$ will reach its smallest value when $y\sqrt{3} - x$ is a minimum. Put

$$y\sqrt{3} - x = p$$

and

$$x\sqrt{3} - y = q.$$

Then

$$p^2 - q^2 = \left(y\sqrt{3} - x\right)^2 - \left(x\sqrt{3} - y\right)^2 = 2\left(y^2 - x^2\right) = 2a^2 ,$$

that is

$$p^2 = 2a^2 + q^2 .$$

p attains its minimum if $q = 0$, that is, if $y = x\sqrt{3}$. This and (2.71) imply that $\tilde{s}$ will reach its smallest value if $x = a\sqrt{2}/2$ and $y = a\sqrt{3/2}$. In other words, among all solids $\tilde{S}$ there is a shape S of minimum surface area

$$s = \tilde{s} - 6a^2\sqrt{3} + 6a\left(\frac{3a}{\sqrt{2}} - \frac{a\sqrt{2}}{2}\right) = \tilde{s} - 6a^2\left(\sqrt{3} - \sqrt{2}\right) .$$

(b) Figure 2.39 shows a face $XYZW$ of a rhombic dodecahedron, constructed from a cube of edge length $XZ = 1$, by the method described on p. 61.

$XYZW$ is a rhombus with diagonals $XZ = 1$ and $WY = \sqrt{2}$.

The diagonals of any of the three rhombuses in the roof of S are of length $PV = 2a\sqrt{\frac{3}{2}}$, and $A'C' = 2a\sqrt{3}$. Thus

$$XZ : WY = 1 : \sqrt{2} = PV : A'C' .$$

This implies that $XYZW$ is similar to $PC'VA'$. Hence the rhombuses in the roof of S have the same angles as the faces of a rhombic dodecahedron.

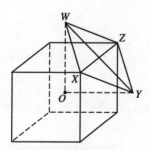

Fig. 2.39

Problem 82

Let $c = ABCDEFGH$ be a cube of edge length 1 (Fig. 2.40). Construct a plane π through the centre O of c, orthogonal to the cube's diagonal HB. The plane π meets the edges EA and GC in their midpoints K and N, the

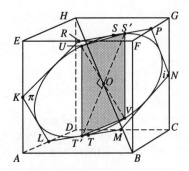

Fig. 2.40

edges *AD* and *FG* in their midpoints *L* and *P*, and the edges *DC* and *EF* in their midpoints *M* and *R*. The polygon $\Pi = KLMNPR$ has congruent edges of length equal to $\sqrt{2}/2$. The distance of the centre *O* of Π from its vertices is also $\sqrt{2}/2$. Thus Π is a regular hexagon. Denote by *i* the circle inscribed in Π. The circle *i* touches *RP* and *LM* at their midpoints *S* and *T* respectively. Choose an arbitrary point *S'* on *i*, near *S* but different from it. Denote by *T'* the point of *i* diametrically opposite to *S'*, and by *U* and *V* the points of *i* such that *S'UT'V* is a square. Since the diagonals *S'T'* and *UV* of *S'UT'V* are of length $\sqrt{2}$, its sides are of length 1. Erect four planes perpendicular to π:

π_1 through *S'U*, π_2 through *UT'*, π_3 through *T'V*, and π_4 through *VS'*. Remove from *c* the region enclosed by the four planes $\pi_1,\ldots,\pi_4$. This creates a hole inside *c*, through which one can push through a cube *c'* congruent to *c*.

Problem 83

Connect the vertex *D* of the cube $c = ABCDEFGH$ to the vertices *E, B,* and *F* (Fig. 2.41). In this way three congruent pyramids are created: Π_1 with base *EFGH* and height *DH*, Π_2 with base *BCGF* and height *DC*, and Π_3 with base *ABFE* and height *DA*. The pyramids Π_1, Π_2, and Π_3 have no

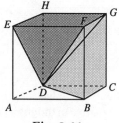

Fig. 2.41

interior points in common and together they form c.. Thus they form a dissection of c into three congruent pyramids.

Problem 84

Let $c = ABCDEFGH$ be a cube of edge length a (Fig. 2.42) and let c' be the image of c under the rotation ρ about BH through an angle $\alpha \leqslant 120°$. Denote the solid forming the common part of c and c' by S.

(i)   Our first am is to determine the position of the point P on the edge FG whose image P' under ρ is on FE. Since $HP' = HP$, triangles HEP' and HGP are congruent. Hence $EP' = GP$, that is $FP' = FP$. The perpendiculars from P and P' onto HB meet at a common point U. The angle $\angle PUP' = \alpha$. Denote the length of PF (and of $P'F$) by x, and the midpoint of PP' by M. Then

$$PM = x\frac{\sqrt{2}}{2}.$$

Hence

$$MU = PM \cot \frac{\alpha}{2} = \frac{x\sqrt{2}}{2} \cot \frac{\alpha}{2}$$

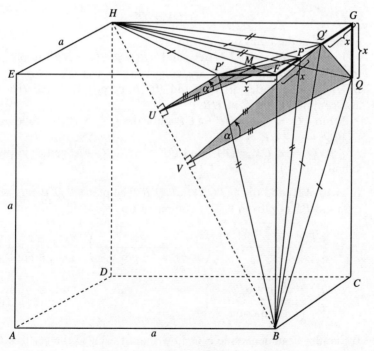

Fig. 2.42

and

$$HM = HF - MF = a\sqrt{2} - \frac{x\sqrt{2}}{2}.$$

HUM and HFB are both right triangles (with right angles at U and F respectively), and they share the angle at H. Therefore they are similar and their corresponding sides are proportional:

$$HM : MU = HB : FB,$$

that is

$$\left(a\sqrt{2} - \frac{x\sqrt{2}}{2}\right) : \frac{x\sqrt{2}}{2}\cot\frac{\alpha}{2} = a\sqrt{3} : a.$$

Hence

$$x = \frac{2a}{1 + \sqrt{3}\cot\alpha/2}.$$

(ii)   Now consider the point Q on GC such that $GQ = x$. Our next task is to show that

$$(*) \qquad\qquad \rho(Q) = Q',$$

where Q' is the point on GF with $GQ' = x$.

Indeed, $HQ' = HQ = BP$ and $BQ' = BQ = HP$ imply that triangles HQB and $HQ'B$ are congruent to triangle BPH (and to $BP'H$). The altitudes of HQB and $HQ'B$, perpendicular to HB, meet at a common point V of HB.

From $VQ' = VQ = UP' = UP$ and from $Q'Q = P'P$, it follows that $\angle QVQ' = \angle PUP' = \alpha$.

$\angle QVQ' = \alpha$, $QV = Q'V$, and $QV \perp BH \perp Q'V$ together imply the validity of $(*)$.

(iii)   In view of (i) and (ii), $\rho(PQB) = P'Q'B$. The plane $P'Q'B$ cuts off the pyramid $BP'Q'F$ from c. The volume of this pyramid is

$$v' = \frac{1}{3}\frac{FQ' \cdot FP'}{2}FB = \frac{1}{6}\left(a - \frac{2a}{1 + \sqrt{3}\cot\alpha/2}\right) \cdot \frac{2a}{1 + \sqrt{3}\cot\alpha/2} \cdot a$$

$$= \frac{a^3}{6} \cdot \frac{2(\sqrt{3}\cot\alpha/2 - 1)}{\left(1 + \sqrt{3}\cot\alpha/2\right)^2}.$$

By similar arguments one can show that at each of the vertices C, D, A, G, and E the image of a rotated face of c cuts off a pyramid congruent to $BP'Q'F$. The solid S is the part of c left after removing these six congruent pyramids.

Thus

$$\text{Volume } S = a^3 - 6 \cdot \frac{a^3}{6} \cdot \frac{2(\sqrt{3}\cot \alpha/2 - 1)}{(1 + \sqrt{3}\cot \alpha/2)^2} = \frac{3a^3(1 + \cot^2 \alpha/2)}{(1 + \sqrt{3}\cot \alpha/2)^2}.$$

If $\alpha = 60°$ then $\cot \alpha/2 = \cot 30° = \sqrt{3}$; in this case

$$\text{Volume } S = \frac{3a^3(1 + 3)}{(1 + 3)^2} = \frac{3a^3}{4}.$$

Problem 85

The regular hexagon $ABCDEF$ with centre O in Fig. 2.43 can be regarded as the orthogonal projection of a cube $c = A'B'C'G'F'H'D'E'$ onto a plane π perpendicular to its diagonal $G'H'$. In this interpretation the points A, B, C, D, E, and F are the orthogonal projections of A', B', C', D', and E' respectively onto π, and O is the orthogonal projection of the diagonal $G'H'$ onto π.

Put $R = P_1P_2 \cap P_3P_4$, $S = P_3P_4 \cap P_5P_6$, and $T = P_5P_6 \cap P_1P_2$. Since, according to the assumption of the problem, $R \in OC$, $s \in OE$ and $T \in OA$, the straight lines RS, ST, and TR, can be considered as the projections of the straight lines along which a plane Σ intersects the faces $C'D'E'G'$, $E'F'A'G'$, and $AB'C'G'$ of c.

P_1, P_6, P_2, and P_3 are the orthogonal projections of the points P_1', P_6', P_2', and P_3' onto π at which Σ intersects the edges $A'B'$, $A'F'$, $B'C'$, and $C'D'$ of

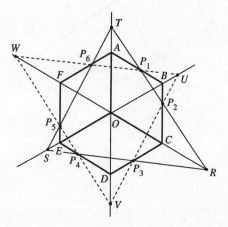

Fig. 2.43

c respectively. Since the lines $P_1'P_6'$ and $P_2'P_3'$ are coplanar and not parallel, they must meet at a common point U'. The lines $P_1'P_6'$ and $P_2'P_3'$ lie in $A'B'H'F'$ and $B'C'D'H'$ respectively. Hence their common point U' belongs to the common line $H'B'$ of these planes. This implies that the projections P_1P_6 and P_2P_3 of $P_1'P_6'$ and $P_2'P_3'$ meet at the projection U of U' which must belong to the projection OB of $H'B'$.

For analogous reasons P_2P_3 and P_4P_5 meet at a point V on OD, and P_4P_5 and P_1P_6 meet at a point W on OF.

Problem 86[1]

Let $T = ABCD$ be an arbitrary tetrahedron.

(a) In space, the locus of points equidistant from two (distinct) points P and Q is the plane through the midpoint of PQ, perpendicular to the straight line PQ.

  Denote by M_1, M_2, M_3, M_4, M_5, and M_6 the midpoints of the edges AB, BC, CA, DB, and DC of T (as shown in Fig. 2.44), and by π_i the plane through M_i, and perpendicular to the edge containing M_i, for $i = 1, 2, \ldots, 6$.

  Since $\angle ABC < 180°$, the planes π_1 and π_2 meet along a straight line ℓ. The line ℓ is perpendicular to the plane ABC, and D does not belong to this plane; hence π_4 intersects ℓ at a point O.

$$\left. \begin{array}{l} O \in \pi_1 \text{ implies that} \quad OA = OB; \\ O \in \pi_2 \quad \text{,,} \qquad \text{,,} \quad OB = OC; \\ O \in \pi_4 \quad \text{,,} \qquad \text{,,} \quad OA = OD. \end{array} \right\} \qquad (2.72)$$

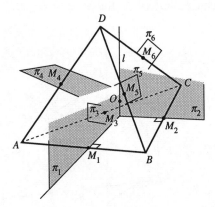

Fig. 2.44

---

[1] Solutions to Problems 86 and 87 can be obtained by using vectors. Here we describe solutions referring only to facts from elementary geometry. Apart from being less sophisticated, this method might provide a better understanding of geometrical configurations.

From (2.72) it follows that

$$OA = OC, \quad \text{i.e.} \quad O \in \pi_3,$$
$$OB = OD, \quad \text{i.e.} \quad O \in \pi_5, \text{ and}$$
$$OC = OD, \quad \text{i.e.} \quad O \in \pi_6.$$

In other words the planes π_i, $i = 1, 2, \ldots, 6$, meet in a common point O. This point is equidistant from all vertices of T; therefore O is the centre of the sphere passing through the vertices of T.

(b) and (c) will be proved together. M_1, M_2, and M_5 are the midpoints of the edges AB, BC, and DB of the tetrahedron $T = ABCD$ (Fig. 2.45). It is well known that AM_5 and DM_1 meet at the centroid G_1 of triangle ABD, and that $AG_1:G_1M_5 = 2:1$. Similarly, CM_5 and DM_2 meet at the centroid G_2 of triangle BCD so that $CG_2:G_2M_5 = 2:1$. The points A, G_2, C, and G_1 belong to the plane AM_5C; hence AG_2 and CG_1 have a common point G.

Since $AG_1:G_1M_5 = CG_2:G_2M_5 = 2:1$ the straight-line segments AC and G_1G_2 are parallel, and $AC = 3G_1G_2$. Thus triangle ACG is an enlargement of triangle G_1G_2G with scale factor 3. This implies that

$$AG:GG_2 = CG:GG_1 = 3:1. \tag{2.73}$$

Denote by G_3 and G_4 the centroids of the faces of CAD and ABC respectively. Using the same arguments as above, one can show that AG_2 and BG_3 must meet at a point G' such that

$$AG':G'G_2 = BG':G'G_3 = 3:1 \tag{2.74}$$

and that AG_2 and DG_4 meet at a point G'' such that

$$AG'':G''G_2 = DG'':G''G_4 = 3:1. \tag{2.75}$$

(2.73), (2.74), and (2.75) together imply that $G = G' = G''$.

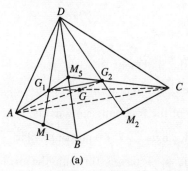

(a)

Fig. 2.45(a)

Thus AG_2, CG_1, BG_3, and DG_4 meet at a common point G that divides each of them in the ratio 3:1. The point G is called the centroid of T.

Remark. The centroid G of T is the midpoint of the straight-line segments joining the midpoints of the opposite edges.

Proof. Consider the triangle AM_2D, where M_2 is the midpoint of BC. mark on DM_2 the centroid G_2 of the face DBC of T. The point G_2 divides DM_2 in the ratio 2:1. Join A to G_2 and mark on AG_2 the centroid G of T. The point G divides AG_2 in the ratio 3:1. Join M_2 to G and denote the intersection of the straight line M_2G with AD by X. let V be the midpoint of AG_2.

From V and G_2 draw parallels to M_2X, which meet AD at Y and Z respectively. The above ratios imply that $XZ = \frac{1}{3}XD$ and $XY = \frac{1}{3}XA$.

Since $XZ = XY$, $XD = XA$. But then $X = M_4$, the midpoint of AD. Further, $XM_2 = \frac{3}{2}ZG_2$ and $ZG_2 = \frac{4}{3}XG$. Hence $XM_2 = \frac{3}{2} \cdot \frac{4}{3}XG = 2XG$. This implies that G is the midpoint of XM_2, that is, of M_4M_2.

By similar arguments one proves that G is the midpoint of M_1M_6 and of M_3M_5.

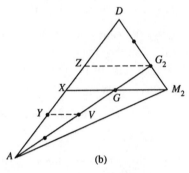

(b)

Fig. 2.45(b)

Problem 87

We shall prove that the statement

(S_1) The tetrahedron T has an orthocentre

is equivalent to each of the following statements:

(S_2) the opposite edges of T are perpendicular;

(S_3) one altitude of T passes through the orthocentre of the corresponding base;

(S₄) the straight-line segments connecting the midpoints of opposite edges are of equal length.

(a) (i)   (S₁) implies (S₂).

Proof. Let H be the orthocentre of the tetrahedron $T = ABCD$. Since the altitudes AA' and DD' of T meet at H (Fig. 2.46(a)), they form a plane AHD.

$$AA' \perp BCD, \quad \text{hence} \quad BC \perp AA';$$

$$DD' \perp ABC, \quad \text{hence} \quad BC \perp DD'.$$

Since BC is perpendicular to two non-parallel lines of the plane AHD, it is perpendicular to the plane AHD.
Hence $BC \perp AD$.

Similar arguments show that $AB \perp DC$ and $AC \perp BD$.

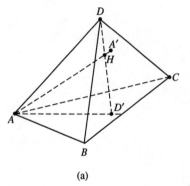

(a)   (b)

Fig. 2.46

(ii)   (S₂) implies (S₁).

Proof. Let DD' be the altitude of T through D. The edge BC is perpendicular to DD' and by assumption to AD. thus BC is perpendicular to the plane ADD'. Therefore the straight line AD' meets BC at right angles at a point D'' (Fig. 2.46(b)).

Drop the perpendicular AA' from A to DD''. Since $AA' \perp DD''$ and $AA' \perp BC$, it follows that AA' is perpendicular to the plane DBC. Thus AA' is the altitude of T through A.

AA' and DD' belong to a common plane and are not parallel; hence they meet in a point H.

By similar arguments one can prove that the altitude BB' meets AA' and DD' at H' and H'' respectively. Since H' and H'' belong to the plane ADD'', and BB' meets ADD'' at a single point, it follows that $H' = H''$. Thus H' and H'' coincide with the unique common point H of AA' and DD'. For a similar reason, the altitude CC' also passes through H. This shows that T is orthocentric.

(S_1) and (S_2) are equivalent statements.

(b) (i) (S_3) implies (S_2).

Proof. Suppose that the altitude AA' of T passes through the orthocentre A' of BCD. This implies that BA', DA', and CA' belong to the altitudes BB'', DD'', and CC'' of triangle BCD (Fig. 2.47). Thus

$$CD \perp BB'' \quad \text{and} \quad CD \perp AA' \quad \text{hence} \quad CD \perp BA.$$

$$DB \perp CC'' \quad \text{and} \quad DB \perp AA' \quad \text{hence} \quad DB \perp CA.$$

$$BC \perp DD'' \quad \text{and} \quad BC \perp AA' \quad \text{hence} \quad BC \perp AD.$$

This proves (i).

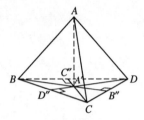

Fig. 2.47

(ii) (S_2) implies (S_3).

Proof. Let AA' be the altitude of T through A.

Since $AA' \perp CD$ and $AB \perp CD$ (by assumption) it follows that $CD \perp BAA'$. Hence $CD \perp BA'$. Thus A' belongs to the altitude BB'' of BCD. Similarly, $AA' \perp BC$ and $AD \perp CB$ (by assumption). Thus $BC \perp ADA'$, and therefore $BC \perp DA'$. Hence A' belongs to the altitude DD'' of BCD. This implies that A' is the orthocentre of BCD.

From $(S_2) \leftrightarrow (S_3)$ and $(S_2) \leftrightarrow (S_1)$ it follows that (S_3) is equivalent to (S_1).

(c) (i) (S_4) implies (S_2).

Proof. Through each edge of T draw a plane, parallel to the opposite edge. The six planes, constructed in this way, form a parallelepiped P, circumscribed about T (Fig. 2.48). The straight-line segments joining the midpoints of the opposite edges of T are congruent to the edges of P. By our assumption, these line-segments are of equal length; hence the faces of P are rhombuses. The diagonals of a rhombus meet at right angles. The

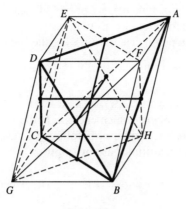

Fig. 2.48

diagonals in the three pairs of parallel faces of P are parallel to the three pairs of opposite edges of T (see Fig. 2.48). This implies that the opposite edges of T are perpendicular.

Thus (S_2) follows from (S_4).

(ii)   The converse of (i) is proved by reversing the arguments used in (c) (i): If the pairs of opposite edges of T are perpendicular, then the faces of the circumscribed parallelepiped P are rhombuses. Thus the straight-line segments joining the midpoints of the opposite edges of T are of equal length. Hence (S_2) implies (S_4).

From $(S_2) \leftrightarrow (S_4)$ and $(S_2) \leftrightarrow (S_1)$ it follows that (S_4) is equivalent to (S_1).

Problem 88

In any tetrahedron the straight-line segments joining the midpoints of the opposite edges meet at the centroid G of the tetrahedron (see Remark on p. 220). In an orthocentric tetrahedron the straight-line segments joining the midpoints of the opposite edges are congruent (Problem 86(c)). Hence if T is an orthocentric tetrahedron, then the centroid G of T is the centre of a sphere S passing through the midpoints of the sides of all faces of T. This implies that S cuts each plane carrying a face of T in the nine-point circle of that face. In other words,

S contains the nine-point circles of all faces of T.

Thus S contains nine distinguished points from each face, namely, (a) the midpoints of the sides, (b) the feet of the altitudes, and (c) the midpoints of the straight-line segments joining the orthocentres of the faces to the vertices.

S carries six points of type (a), six points of type (b) (because the feet of the altitudes perpendicular to a common edge of T coincide), and $4 \cdot 3 = 12$ points of type (c). Therefore S is called the 24-point sphere.

Problem 89

Let T be a tetrahedron with faces f_i of area a_i for $i = 1, 2, 3, 4$. If a sphere S of radius r touches all planes which enclose T, then the volume of T can be expressed in terms of the volumes of four pyramids, whose bases are the faces f_i of T and whose altitudes are of length r:

$$\text{Volume } T = \frac{r}{3}(\varepsilon_1 a_1 + \varepsilon_2 a_2 + \varepsilon_3 a_3 + \varepsilon_4 a_4), \quad \text{where} \quad \varepsilon_i \in \{1, -1\}.$$

There are $2^4 = 16$ combinations of $\varepsilon_i \in \{1, -1\}$ leading to 16 corresponding expressions $(r/3) \sum_{i=1}^{4} \varepsilon_i a_i$. Not all of them represent the volume of T since the volume of T must be positive. In fact, if for a certain combination $\varepsilon_1^*, \varepsilon_2^*, \varepsilon_3^*, \varepsilon_4^*$ of the $\varepsilon_i \in \{1, -1\}$

$$\frac{r}{3}(\varepsilon_1^* a_1 + \varepsilon_2^* a_2 + \varepsilon_3^* a_3 + \varepsilon_4^* a_4) = \text{Volume } T,$$

then the combination $-\varepsilon_1^*, -\varepsilon_2^*, -\varepsilon_3^*, -\varepsilon_4^*$ leads to the negative number

$$-\frac{r}{3}(\varepsilon_1^* a_1 + \varepsilon_2^* a_2 + \varepsilon_3^* a_3 + \varepsilon_4^* a_4),$$

which cannot be the volume of T.

Thus, at most $16/2 = 8$ combinations of the ε_i yield positive values for $(r/3) \sum_{i=1}^{4} \varepsilon_i a_i$, that is, there can be at most eight different spheres touching the planes of all faces of an arbitrary tetrahedron.

For any tetrahedron there exist at least five spheres touching the planes which carry the faces f_i: one sphere inside T, touching all faces of T, and four spheres outside T, each of them touching exactly one of the faces f_i. If T is a regular tetrahedron, then the expression $(r/3) \sum_{i=1}^{4} \varepsilon_i a_i$ is positive for exactly 5 combinations of ε_i: $(1, 1, 1, 1)$, $(-1, 1, 1, 1)$, $(1, -1, 1, 1)$, $(1, 1, -1, 1)$, and $(1, 1, 1, -1)$. Thus there exist exactly five spheres touching all planes which enclose a regular tetrahedron.

Problem 90

Let $T = ABCD$ be an arbitrary tetrahedron. Denote by $\sigma_1, \sigma_2, \ldots, \sigma_6$ the planes through the midpoints of the edges AB, BC, CA, AD, BD, and CD respectively, each of them perpendicular to the opposite edge of T (Fig. 2.49). Denote by O the centre of the sphere about T, and by G the centroid of T.

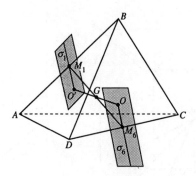

Fig. 2.49

We shall prove that σ_i, $i = 1, \ldots, 6$, passes through the point O' symmetric to O with respect to G.

Proof. Let M_1 be the midpoint of AB and M_6 the midpoint of CD. The centroid G of T is the midpoint of $M_1 M_6$ (see Remark on p. 220). $GO = GO'$ and $GM_6 = GM_1$ imply that OM_6 and $O'M_1$ are parallel. Since $OM_6 \perp DC$, it follows that $O'M_1 \perp DC$. Hence $O'M_1$ is contained in the plane σ_1.

Using similar arguments one can prove that O' is a common point of all the planes $\sigma_1, \sigma_2, \ldots, \sigma_6$.

Problem 91

Figure 2.50 shows the pyramids $P = ABCDV$ and $Q = A'B'C'D'V'$ with square bases of edge length a and altitudes VO and $V'O'$ of length $a/2$. The foot O of VO is the centre of $ABCD$, and the foot O' of $V'O'$ is the midpoint of $D'C'$. The edges AB, BC, CD, DA, VA, VB, VC, and VD of P are numbered ① to ⑧ (see Fig. 2.50(a)), and the edges of Q are numbered in a similar fashion (Fig. 2.50(b)).

Denote the dihedral angle of P with edge ① by α_i and the dihedral angle of Q with edge ① by β_i, $i = 1, 2, \ldots, 8$.

Our first task is to determine the angles α_i and B_i. We have

(i)
$$\alpha_1 = \alpha_2 = \alpha_3 = \alpha_4 = \frac{\pi}{4}.$$

Proof. These angles are congruent to $\angle VMO$.

(ii)
$$\beta_4 = \beta_2 = \frac{\pi}{4}.$$

Proof. These angles are congruent to $V'D'O'$.

(iii)
$$\beta_7 = \beta_8 = \beta_3 = \frac{\pi}{2}.$$

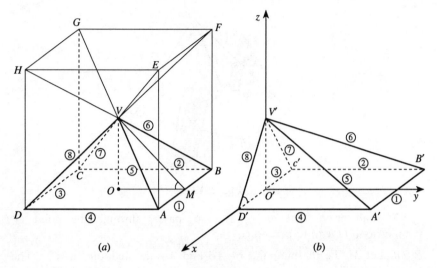

Fig. 2.50

(iv) $$\sin \beta_1 = \frac{\sqrt{5}}{5} \quad \text{and} \quad \cos \beta_1 = \frac{2\sqrt{5}}{5}.$$

(v) $$\alpha_5 = \alpha_6 = \alpha_7 = \alpha_8 = \frac{2\pi}{3}$$

Proof. (v) can be shown by completing P to the cube $ABCDEFGH$, as shown in Fig. 2.50(a). In this cube the edge VA is shared by the three congruent pyramids P, $ABFEV$, and $AEHDV$. Their dihedral angles with edge VA add up to 2π, hence $\alpha_5 = 2\pi/3$.

(vi) $$\beta_5 = \beta_6, \quad \cos B_5 = -\frac{\sqrt{10}}{5} \quad \text{and} \quad \sin \beta_5 = \frac{\sqrt{15}}{5}.$$

Proof. These results can be obtained by introducing a cartesian coordinate system with origin O', in which the points A', B', D', and V' have coordinates $(a/2, a, 0)$, $(-a/2, a, 0)$, $(a/2, 0, 0)$, and $(0, 0, a/2)$ respectively. In this coordinate system the planes $A'V'D'$ and $A'V'B'$, forming the dihedral angle β_5, have equations $x + z = a/2$ and $y + 2z = a$. Thus

$$\cos \beta_5 = -\frac{2}{\sqrt{2} \cdot \sqrt{5}} = -\frac{\sqrt{10}}{5},$$

$$\sin \beta_5 = \sqrt{1 - \tfrac{10}{25}} = \tfrac{\sqrt{15}}{5}.$$

The same holds for $\cos \beta_6$ and $\sin \beta_6$.

The next step is to assume that α_i and β_i satisfy Dehn's condition (C) (this assumption will lead to contradiction).

Suppose that there exist non-negative integers $a_1, \ldots, a_8$ and $b_1, \ldots, b_8$ and an integer k such that

$$a_1 \frac{\pi}{4} + a_2 \frac{\pi}{4} + a_3 \frac{\pi}{4} + a_4 \frac{\pi}{4} + a_5 \frac{2\pi}{3} + a_6 \frac{2\pi}{3} + a_7 \frac{2\pi}{3} + a_8 \frac{2\pi}{3}$$

$$-\left(b_1\beta_1 + b_2 \frac{\pi}{4} + b_3 \frac{\pi}{2} + b_4 \frac{\pi}{4} + b_5\beta_5 + b_6\beta_6 + b_7 \frac{\pi}{2} + b_8 \frac{\pi}{2}\right) = k\pi.$$

This condition reduces to

$$n \frac{\pi}{4} + m \frac{2\pi}{3} - \ell \frac{\pi}{2} - b_1\beta_1 - (b_5 + b_6)\beta_5 = k\pi,$$

or, after multiplying by 12, to

$$r\beta_1 + s\beta_5 = t\pi,$$

where r and s are even natural numbers, and t is an integer. Hence

$$\sin r\beta_1 = \pm \sin s\beta_5. \tag{2.76}$$

According to a well-known formula (see the remark that follows the solution of this problem),

$$\sin r\beta_1 = \sum_{k=1}^{\left[\frac{r+1}{2}\right]} (-1)^{k-1} \binom{r}{2k-1} (\cos \beta_1)^{r-2k+1} (\sin \beta_1)^{2k-1}.$$

In view of (iv), this sum is equal to

$$\sin r\beta_1 = \sum_{k=1}^{\left[\frac{r+1}{2}\right]} (-1)^{k-1} \binom{r}{2k-1} \left(\frac{2\sqrt{5}}{5}\right)^{r-2k+1} \left(\frac{\sqrt{5}}{5}\right)^{2k-1}$$

$$= \sum_{k=1}^{\left[\frac{r+1}{2}\right]} (-1)^{k-1} \binom{r}{2k-1} 2^{r-2k+1} \frac{(\sqrt{5})^r}{5^r}.$$

Since r is even, $(\sqrt{5})^r$ is a power of 5 with integer exponent; thus

$$\sin r\beta_1 \text{ is a rational number.} \tag{2.77}$$

On the other hand, according to (vi)

$$\sin s\beta_5 = \sum_{k=1}^{\left[\frac{s+1}{2}\right]} (-1)^{k-1} \binom{s}{2k-1} (\cos \beta_5)^{s-2k+1} (\sin \beta_5)^{2k-1}.$$

$$= \sum_{k=1}^{\left[\frac{s+1}{2}\right]} (-1)^{k-1} \binom{s}{2k-1} \left(-\frac{\sqrt{10}}{5}\right)^{s-2k+1} \left(\frac{\sqrt{15}}{5}\right)^{2k-1}$$

$$= \sum_{k=1}^{\left[\frac{s+1}{2}\right]} \binom{s}{2k-1} \frac{(-1)^k}{5^s} \left(\sqrt{10}\right)^s \left(\frac{3}{2}\right)^k \sqrt{\frac{2}{3}}.$$

This means (since s is even) that

$$\sin s\beta_5 \text{ is an irrational number.} \qquad (2.78)$$

(2.77) and (2.78), together with (2.76), yield a contradiction, since no rational number is equal to an irrational number. Thus P and Q are not equidecomposable.

Remark. Let n be a natural number and α an arbitrary angle. The nth power of the complex number $\cos \alpha + i \sin \alpha$ can be evaluated in two different ways:

According to the binomial formula

$$(\cos \alpha + i \sin \alpha)^n = \sum_{k=0}^{n} \binom{n}{k} (\cos \alpha)^{n-k} (i \sin \alpha)^k.$$

By applying the theorem of De Moivre (1667–1754),

$$(\cos \alpha + i \sin \alpha)^n = \cos n\alpha + i \sin n\alpha.$$

Thus

$$\cos n\alpha + i \sin n\alpha = \sum_{k=0}^{n} \binom{n}{k} (\cos \alpha)^{n-k} (i \sin \alpha)^k.$$

By equating the imaginary parts of the expressions on both sides of this identity we find that

$$\sin n\alpha = \sum_{k=1}^{\left[\frac{n+1}{2}\right]} (-1)^{k-1} \binom{n}{2k-1} (\cos \alpha)^{n-2k+1} (\sin \alpha)^{2k-1}.$$

Problem 92

The dihedral angles of a cube C are $\alpha_i = \pi/2$, $i = 1, \ldots, 12$, and the dihedral angles of a regular tetrahedron T are β_i, such that $\cos \beta_i = \frac{1}{3}$ and $\sin \beta_i = 2\sqrt{2}/3$, $i = 1, \ldots, 6$.

Suppose that α_i and β_i satisfy Dehn's condition ($\mathbb{C}$). This implies that

$$12 \frac{\pi}{2} + 6\beta_i = k\pi$$

for some integer k, that is

$$6\beta_i = (k - 6)\pi.$$

From the last equation it follows that

$$\sin 6\beta_i = 0.$$

However,

$$\sin 6\beta_i = \binom{6}{1} \cos^5 \beta_i \sin \beta_i - \binom{6}{3} \cos^3 \beta_i \sin^3 \beta_i + \binom{6}{5} \cos \beta_i \sin^5 \beta_i$$

$$= 6 \cdot \left(\frac{1}{3}\right)^5 \frac{2\sqrt{2}}{3} - 20 \cdot \left(\frac{1}{3}\right)^3 \left(\frac{2\sqrt{2}}{3}\right)^3 + 6 \cdot \frac{1}{3} \left(\frac{2\sqrt{2}}{3}\right)^5 \neq 0.$$

Thus α_i and β_i do not satisfy Dehn's condition. Therefore C and T are not equidecomposable.

Problem 93

$P = A_6 B_6 C_6 D_6 V$ is a pyramid with square base $A_6 B_6 C_6 D_6$ of side length a and height VO of length $a/2$, where O is the centre of $ABCD$. We shall divide P into parts, from which a cuboid C with the same base as P can be assembled, as follows.

Divide VO into six equal parts and through the division points draw planes to $ABCD$. They intersect P along squares $A_i B_i C_i D_i$, $i = 1, 2, \ldots, 5$, as shown in Fig. 2.51. Denote by Π_i the part of P enclosed between $A_i B_i C_i D_i$ and $A_{i+1} B_{i+1} C_{i+1} D_{i+1}$, $i = 1, \ldots, 5$. Through each of the edges $A_i B_i, B_i C_i, C_i D_i$, and $D_i A_i$ of Π_i pass a plane, perpendicular to $A_{i+1} B_{i+1} C_{i+1} D_{i+1}$. In this way Π_i is divided into nine parts: a cuboid $\overline{C}_i$ of dimensions $i \cdot (a/6)$, $i \cdot (a/6)$, $a/12$, four congruent triangular prisms, which can be assembled into a cuboid $\overline{C}_i'$ of dimensions $i \cdot (a/6)$, $a/6$, $a/12$, and four pyramids which fit together into a pyramid $\overline{P}_i$, congruent to $A_1 B_1 C_1 D_1 V$ (Fig. 2.52).

We shall now cut the cuboids $\overline{C}_i$ and $\overline{C}_i'$, $i = 1, 2, \ldots, 5$, into cuboids C^* of dimensions $a/6$, $a/6$, $a/12$. This gives altogether $(1 + 4 + 9 + 16 + 25) + (1 + 2 + 3 + 4 + 5) = 55 + 15 = 70$ cuboids C^*. The pyramids $P_1, \ldots, P_6$ fit together into a cube of edge length $a/6$ which we cut into two cuboids C^*.

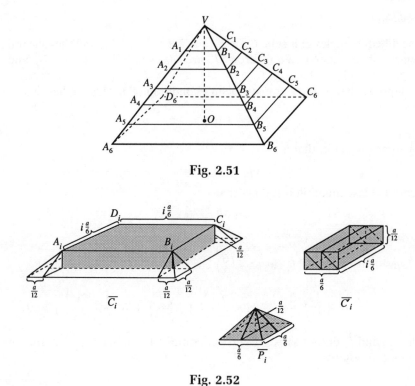

Fig. 2.51

Fig. 2.52

From the 72 cuboids C^* we can construct a cuboid C of dimensions a, a, and $a/6$. The cuboid C has the same base as P; its volume is the same as the volume of P.

We have proved that P and C are equidecomposable.

Problem 94

We shall construct a set S of infinitely many congruent spheres, each of them touching 12 spheres of S, as follows.

We tessellate three-dimensional space by means of congruent cubes, and paint each cube white or grey, so that cubes with a common face have different colours. We draw the 'midspheres', that is the spheres through the midpoints of the cube's edges, of each white cube. In this way we obtain an infinite set S of congruent spheres. Two spheres s_i and s_j of S have a non-empty intersection if and only if the white cubes c_i and c_j containing them as midspheres have a common edge; in that case s_i and s_j touch one another at the midpoint of the edge which belongs to both c_i and c_j (Fig. 2.53).

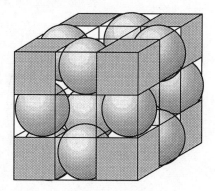

Fig. 2.53

It is easy to see that an arbitrary white cube c^* is surrounded by 12 cubes $c_1, \ldots, c_{12}$ of the tessellation, each of them sharing an edge with c^*. Hence the midspheres s^* of c^* touch each of the midspheres $s_1, \ldots, s_{12}$ of $c_1, \ldots, c_{12}$ respectively.

Remark. The planes, tangent to s^* at the midpoints of the edges of c^*, enclose a rhombic dodecahedron (see p. 61); thus the rhombic dodecahedra circumscribed about the spheres of S tessellate (three-dimensional) space.

Problem 95

Shortest curves on surfaces, joining two points of the surface, are called *geodesics*. A rigorous study of geodesics. can be pursued with the help of differential geometry. The following solution of Problem 95 is described in Sharygin [29].

Let S be a sphere with centre O, and let P and Q be two arbitrary, distinct points on S. The points P and Q divide the great circle of S on which they lie into two arcs, a' and a''. Suppose that $a' \leqslant a''$. Our aim is to prove that a' is the shortest path on the surface of S which connects P and Q (Fig. 2.54).

Suppose that the shortest path on S connecting P and Q is a curve b, different from a'. In this case there exists a point R on a' which does not belong to b. From R drop perpendiculars RP' and RQ' to OP and OQ respectively. The plane through P', perpendicular to OP, meets S in a circle p, and the plane through Q', perpendicular to OQ, meets S in a circle q. The circles p and q have only the point R in common. p and q intersect b at two different points M and N respectively. Thus b is divided into three parts: b_1 between P and M, b_2 between M and N, and b_3 between N and Q.

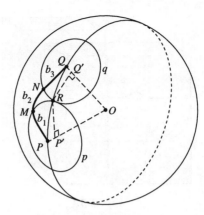

Fig. 2.54

Now rotate the spherical cap, bounded by p and containing P, about OP so that M is moved into the position of R. Under this rotation b_1 is mapped onto a curve b_1', congruent to b, with endpoints P and R. Similarly, rotate the spherical cap, bounded by q and containing Q, about OQ so that N is moved into the position of R. This rotation maps b_3 onto b_3', congruent to b_3, with endpoints R and Q.

The length of the curve $b_1' + b_3'$ is equal to the length of $b_1 + b_3$ which is shorter than $b_1 + b_2 + b_3 = b$. This contradicts our assumption that b is the shortest path on S connecting P to Q. Hence any point R of a' belongs to b. In other words, the arc a' is the shortest path on S between P and Q.

Problem 96

Let us assume that there are three arcs a_1, a_2, a_3 of three great circles of a sphere S, such that each has a central angle of $300°$, and that no two of them have a point in common. Denote by α_i the plane containing a_i, $i = 1, 2, 3$.

α_1 and α_2 meet along a straight line which contains a diameter AA' of S; α_2 and α_3 meet along a straight line containing a diameter BB' of S; and the intersection of α_3 and α_1 is a straight line containing a diameter CC' of S (Fig. 2.55).

Since the central angle of a_1 is greater than $180°$, a_1 must contain at least one of the diametrically opposite points A, A' and at least one of the diametrically opposite points C and C'; suppose, for definiteness, that a_1 contains A' and C'.

Similarly, since the central angle of a_2 is greater than $180°$, a_2 must contain at least one of A, A' and at least one of B, B'. According to our assumptions, a_1 and a_2 have no points in common and a_1 contains A'. This implies that a_2 contains A; suppose, for definiteness, that B' is on a_2.

Similar arguments imply that a_3 contains B and C.

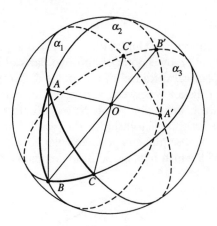

Fig. 2.55

Denote by O the centre of S.

Since a_1 contains A', C' but not A and C, it follows that

$$360° - \angle AOC > 300°.$$

Since a_2 contains A and B' but not A' and B,

$$360° - \angle A'OB > 300°.$$

Finally, since a_3 contains B and C but not B' and C',

$$360° - \angle B'OC' > 300°.$$

In view of the above inequalities

$$\angle AOC < 60°,$$
$$\angle BOC = \angle B'OC' < 60°$$

and

$$\angle AOB = 180° - \angle A'OB > 120°.$$

Hence

$$\angle AOB > \angle AOC + \angle BOC.$$

This is a contradiction, since in the tetrahedron $OABC$ the angle $\angle AOB$ between the edges OA and OB is less than the sum of the angles $\angle AOC$ and $\angle BOC$, between the edges OA, OC and between OC and OB respectively (verify this).

Therefore, the assumption that no two of the three arcs a_1, a_2, a_3 have a point in common is false.

Problem 97

Let S be a sphere with centre O and radius $r = 1$. p is a 'polygonal line' on S of length $\ell < \pi$. Divide p by a point P into two parts of equal length. This implies thai the 'spherical distance' of any point $X \in p$ from P is less than $\pi/2$. (The spherical distance between two points of S is the length of the shortest route between these points, on S; see Problem 95). The locus of the points Y of S with spherical distance $PY = \pi/2$ is the set of the points of the great circle c whose plane is perpendicular to PO (Fig. 2.56). Hence c is a great circle of S which h as no point in common with p.

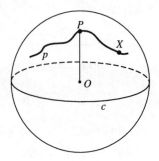

Fig. 2.56

Chapter VIII

Problem 98

Let $P(x_1, x_2, \ldots, x_n)$ be a given point of Σ_n, and let $P'(x_1', x_2', \ldots, x_{i-1}', 0, x_{i+1}', \ldots, x_n')$ be any point of the hyperplane Π_i of Σ_n with equation $x_i = 0$. The distance between P and P' is given by the formula

$$PP' =$$

$$\sqrt{(x_1 - x_1')^2 + \ldots + (x_{i-1} - x_{i-1}')^2 + x_i^2 + (x_{i+1} - x_{i+1}')^2 + \ldots + (x_n - x_n')^2}.$$

The above expression reaches its minimum for $x_j - x_j' = 0$ for all $j = 1, 2, \ldots, i-1, i+1, \ldots, n$. In this case $PP' = \sqrt{x_i^2} = |x_i|$.
  Thus $d(P, \Pi_i) = |x_i|$.

Problem 99

(a) $\angle POP' = 90°$, if and only if POP' is a right triangle. Then

$$P'P^2 = PO^2 + P'O^2,$$

that is

$$\left(\sqrt{(x_1 - x_1')^2 + (x_2 - x_2')^2 + \ldots + (x_n - x_n')^2} \right)^2$$

$$= \left(\sqrt{x_1^2 + x_2^2 + \ldots + x_n^2} \right)^2 + \left(\sqrt{(x_1')^2 + (x_2')^2 + \ldots + (x_n')^2} \right)^2.$$

This equation is satisfied if and only if

$$x_1 x_1' + x_2 x_2' + \ldots + x_n x_n' = 0.$$

(b) Let $\angle POP' = \alpha$, where $0° < \alpha < 180°$. Then, according to the cosine rule applied to triangle POP',

$$P'P^2 = PO^2 + P'O^2 - 2PO \cdot P'O \cos \alpha.$$

235

Hence

$$\cos \alpha = \frac{PO^2 + P'O^2 - P'P^2}{2PO \cdot PO'}.$$

The above equation leads to

$$\cos \alpha = \frac{x_1 x_1' + x_2 x_2' + \ldots + x_n x_n'}{\sqrt{x_1^2 + x_2^2 + \ldots + x_n^2} \cdot \sqrt{(x_1')^2 + (x_2')^2 + \ldots + (x_n')^2}}. \qquad (2.79)$$

(2.79) can be extended to the cases when $\alpha = 0°$ or $\alpha = 180°$. In these cases $x_i' = kx_i$ for all $i = 1, \ldots, n$; here k is a constant which is positive if $\alpha = 0°$ and negative if $\alpha = 180°$. This implies that

$$\cos \alpha = \frac{k(x_1^2 + \ldots + x_n^2)}{|k|(x_1^2 + \ldots + x_n^2)} = \frac{k}{|k|},$$

yielding the well known values $\cos 0° = 1$ and $\cos 180° = -1$.

Problem 100

(a) Denote by P_i the point on Ox_i with coordinates $\bar{x}_j = 0$ for $j \neq i$ and $\bar{x}_j = 1$.

   (i) Let ℓ be an arbitrary line of Π_i, passing through the origin $0(0, 0, \ldots, 0)$. Denote the coordinates of a fixed point $P \neq O$ on ℓ by $a_1, \ldots, a_n$. Since P belongs to Π_i, $a_i = 0$.

$$a_1 \bar{x}_1 + a_2 \bar{x}_2 + \ldots + a_{i-1} \bar{x}_{i-1} + a_i \bar{x}_i + a_{i+1} \bar{x}_{i+1} + \ldots + a_n \bar{x}_n$$
$$= a_1 \cdot 0 + a_2 \cdot 0 + \ldots + a_{i-1} \cdot 0 + 0 \cdot 1 + a_{i+1} \cdot 0 + \ldots + a_n \cdot 0 = 0.$$

Hence, $\angle P_i OP = 90°$ (see Problem 99(a)).

   This shows that Ox_i is orthogonal to all lines of Π_i, passing through the origin of Σ.

   (ii) Let ℓ' be a line of Π_i which does not contain O. In that case, ℓ' and Ox_i belong to a unique three-dimensional subspace Σ_3 of Σ_n. The subspace Σ_3 contains a unique line ℓ, parallel to ℓ' and passing through O. The line ℓ belongs to Π_i; hence, according to (i), Ox_i is orthogonal to ℓ. Thus in view of (∗) (on p. 75), Ox_i is orthogonal to ℓ'.
   From (∗ ∗) (on p. 75) it follows that Ox_i is orthogonal to Π_i.

(b) If $j \neq i$ then Π_j contains the line Ox_i. According to (a) the line Ox_i is orthogonal to Π_i. Hence, in view of (∗ ∗ ∗) (p. 75), Π_j is orthogonal to Π_i for $j \in \{1, \ldots, n\}, j \neq i$.

Problem 101

(a) and (b) will be discussed together.

Consider the n-dimensional hypercube $C_n(a)$, bounded by the hyperplanes $x_k = 0$ and $x_k = a$ for $k = 1, \ldots, n$. Denote by $B_n(a)$ the 'base' of $C_n(a)$, that is, its $(n-1)$-dimensional part contained in the hyperplane $x_n = 0$.

Let τ be the translation mapping the point $(x_1, \ldots, x_{n-1}, x_n)$ of Σ_n onto the point $(x_1, x_2, \ldots, x_{n-1}, x_n + a)$. The image of $B_n(a)$ under τ is the $(n-1)$-dimensional part of $C_n(a)$ contained in the hyperplane $x_n = a$; call this part $B'_n(a)$. It is easy to verify that:

(i)   the image of any j-dimensional part $B_j(a)$ of $B_n(a)$ under τ is a j-dimensional part $B'_j(a)$ of $B'_n(a)$;

(ii)   the vertices of $B_j(a)$ and of $B'_j(a)$ together form the set of vertices of a $(j+1)$-dimensional part of $C_n(a)$ (shown in Fig. 2.57 for $C_2(a)$ and $C_3(a)$).

(iii) any $(j+1)$-dimensional part of $C_n(a)$ which does not belong to $B_n(a)$ or to $B'_n(a)$ meets $B_n(a)$ and $B'_n(a)$ in j-dimensional parts $B_j(a)$ and $B'_j(a)$ respectively, where $B'_j(a)$ is the translate of $B_j(a)$ under τ.

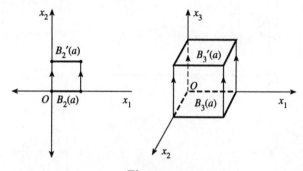

Fig. 2.57

The above arguments lead to a recursion formula for the number $p_{n,i}$ of the i-dimensional parts of $C_n(a)$:

$$p_{n,i} = 2p_{n-1,i} + p_{n-1,i-1} \quad \text{for} \quad i = 1, \ldots, n-1 \tag{2.80}$$

The zero-dimensional parts of $C_n(a)$ are the zero-dimensional parts of $B_n(a)$ and of $B'_n(a)$. Thus

$$p_{n,0} = 2p_{n-1,0}. \tag{2.81}$$

Finally,

$$p_{n,n} = 1. \tag{2.82}$$

Using (2.80) and (2.81) one can prove by induction that

$$p_{n,i} = \binom{n}{i} 2^{n-i} \quad \text{for} \quad i \geq 0, \, n \geq i. \tag{2.83}$$

Here we shall provide a combinatorial proof deducing (2.83) from (2.80) and (2.81) based on a combinatorial interpretation of the numbers $p_{n,i}$ in the study of a network. The advantage of this proof is that it explains the reason for the appearance of the binomial coefficient $D\binom{n}{i}$ in the formula.

Figure 2.58 shows a network of roads. In the top row of the network each node, apart from that at O, can be approached only from the node on its left, along a single path. In the 0th column each node, other than the node at O, can be approached only from the node above it by one of two prescribed paths. Any other node of the network can be reached from the node at its left by one path, and from the node above it by one of two possible paths.

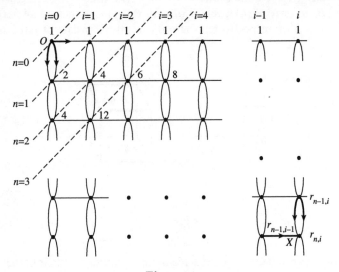

Fig. 2.58

Let $r_{n,i}$ be the number of paths leading from O to the node X in the ith column and on the nth diagonal of the network (see Fig. 2.58). The numbers $r_{n,i}$ satisfy the same recursive relation as the numbers $p_{n,i}$:

$$r_{n,i} = 2r_{n-1,i} + r_{n-1,i-1}.$$

Moreover, the initial value $r_{n,0}$ is the same as $p_{n,0}$

$$\text{Thus} \quad r_{n,i} = p_{n,i} \quad \text{for} \quad n \geq 0, \, i \geq 0, \, n \geq i.$$

The number of paths connecting O with X can be determined as follows.

Each path from O to X consists of n steps, where each step is the part of the path which connects two neighbouring blocks. Of these n steps, i are 'horizontal' and the remaining $n - i$ are 'vertical'. Thus each path from O to X corresponds to a choice of i horizontal steps from the total of n steps; the number of such choices is $\binom{n}{i}$. For any particular choice of i horizontal steps, each of the $n - i$ vertical steps can be taken in two different ways: either along the left 'curved path' ($\subsetneq$), or along the right curved path ($\supsetneq$). Hence

$$p_{n,i} = \binom{n}{i} \cdot \underbrace{2 \cdot 2 \cdot \ldots \cdot 2}_{n-i} = \binom{n}{i} 2^{n-i}.$$

Problem 102

A net of $C_4(a)$ is shown in Fig. 2.59. When the net is stuck together to form the four-dimensional solid $C_4(a)$, the vertices labelled with the same letters (e.g. I_1, I_2, and I_3) must meet at a single point.

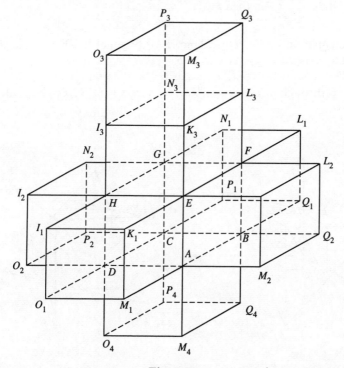

Fig. 2.59

Problem 103

(a) The equation of the plane π, orthogonal to the diagonal $D_3 = \langle O, (a, a, a) \rangle$ of $C_3(a)$ is $x_1 + x_2 + x_3 = p$. In this equation the constant p depends on the position of the point $P = D_3 \cap \pi$.

π intersects the coordinate axis Ox_i at the point X_i, $i = 1, 2, 3$. Moreover, π intersects the line ℓ_i passing through (a, a, a) and parallel to Ox_i at X_i', $i = 1, 2, 3$.

The points $X_1(p, 0, 0)$, $X_2(0, p, 0)$, and $X_3(0, 0, p)$ form an equilateral triangle $\triangle$ (which degenerates into a point if $p = 0$). Similarly, the points $X_1'(p - 2a, a, a)$, $X_2'(a, p - 2a, a)$, and $X_3'(a, a, p - 2a)$ form an equilateral triangle $\triangle'$ ($\triangle'$ is a point if $p = 3a$).

The intersection of the cube $C_3(a)$ with π can be determined by studying the positions of $\triangle$ and $\triangle'$ for various values of p.

(1) If $0 \leqslant p \leqslant a$, then $\triangle \cap \triangle' = \emptyset$; $C_3(a) \cap \pi = \triangle$ (see Fig. 2.60(a)).

(2) If $a < p < 2a$, then $\triangle \cap \triangle'$ is a hexagon H; $C_3(a) \cap \pi = H$ (see Fig. 2.60(b)).

(3) If $2a \leqslant p \leqslant 3a$, then $\triangle \cap \triangle' = \emptyset$; $C_3(a) \cap \pi = \triangle'$.

(b) The equation of the hyperplane Π, orthogonal to $D_4 = \langle O, (a, a, a, a) \rangle$ of $C_4(a)$, is $x_1 + x_2 + x_3 + x_4 = p$.

Π intersects the coordinate axis Ox_i at the point X_i, $i = 1, 2, 3, 4$. The line ℓ_i through (a, a, a, a), $\ell_i \parallel Ox_i$ intersects Π at X_i', $i = 1, 2, 3, 4$.

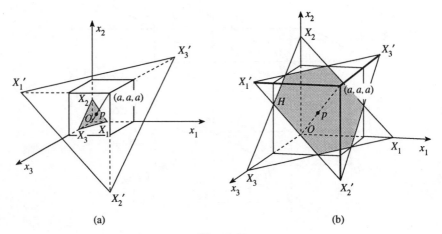

(a)   (b)

Fig. 2.60

The points $X_1(p,0,0,0)$, $X_2(0,p,0,0)$, $X_3(0,0,p,0)$ and $X_4(0,0,0,p)$ form a regular tetrahedron T (which degenerates to a point if $p=0$). The points $X_1'(p-3a,a,a,a)$, $X_2'(a,p-3a,a,a)$, $X_3'(a,a,p-3a,a)$, and $X_4'(a,a,a,p-3a)$ are the vertices of a regular tetrahedron T' (T' is a point if $p=4a$). The intersection of $C_4(a)$ with Π depends on the mutual position of T and T'. Specifically,

(1) if $0 \leqslant p \leqslant a$, then $T \cap T' = \emptyset$; $C_4(a) \cap \Pi = T$,

(2) if $a < p < 3a$, then T and T' have a non-empty intersection K, in this case $C_4(a) \cap \Pi = K$;

(3) if $3a \leqslant p \leqslant 4a$, then $T \cap T' = \emptyset$, $C_4(a) \cap \Pi = T'$.

In the special case when Π passes through the midpoint of D_4, $p = 2a$. Both tetrahedra T and T' contain the points $A_1(a,a,0,0)$, $A_2(a,0,a,0)$, $A_3(a,0,0,a)$, $A_4(0,a,a,0)$, and $A_6(0,0,a,a)$ as the midpoints of their edges. The points A_i are the vertices of a regular octahedron K^*. It is easy to see that K^* is the intersection of T and T'. Hence, if $p = 2a$, then the hyperplane Π intersects the hypercube $C_4(a)$ in a regular octagon.

Problem 104

(i) The shape of the projection of a square or a cube from a centre P onto Σ_1 or Σ_2 depends on the mutual positions of these objects. In Fig. 2.61(a) the square $ABCD$ is projected from a point P on the perpendicular bisector of AB onto a line Σ_1 parallel to AB. P is outside $ABCD$. In this case the projection $A'B'C'D'$ is a straight-line segment. In our diagram the projection of $ABCD$ is the projection $D'C'$ of its edge DC; $D'C'$ consists of the projections $D'A'$, $A'B'$, and $B'C'$ of the three remaining edges of the square.

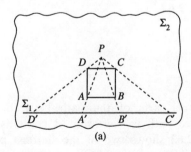
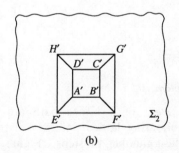

(a)   (b)

Fig. 2.61

Figure 2.61(b) depicts the projection of the cube $C_3 = ABCDEFGH$ from a point P, collinear with the centres of the faces $ABCD$ and $EFGH$, onto $\Sigma_2 \parallel ABCD$. P is outside the cube. The projection of C_3 is a square. In our diagram this square is the projection $E'F'G'H'$ of $EFGH$ onto Σ_2; it consists of five quadrilaterals, which are the projections of the five remaining faces of C_3.

(ii) In Σ_4, the central projection of any point Q from a point P, called the centre, onto a hyperplane $\Sigma_3 \not\ni P$ is the intersection Q' of the straight line PQ with Σ_3 (provided that such an intersection exists).

Let C_4 be a four-dimensional hypercube in Σ_4, P a point on the straight line through the centres of two opposite facets of C_4, and Σ_3 a hyperplane parallel to these facets. If P is outside the hypercube, then it is not difficult to verify (e.g. by using coordinates) that the projection of C_4 is a cube C'_3. C'_3 is the projection of a facet of C_4; it consists of seven solids which are the projections of the remaining seven facets of C_4. Figure 2.62 shows such a central projection of C_4 with vertices $A_1, A_2, \ldots, A_{16}$. The facets of C_4 parallel to Σ_3 are $A_1 A_2 \ldots A_8$ and $A_9 A_{10} \ldots A_{16}$.

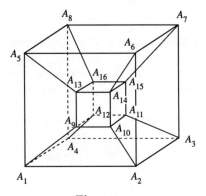

Fig. 2.62

Problem 105

The *steps* of a lattice path are the parts of the path between two neighbouring lattice points. A lattice path is *shortest* if it contains the smallest number of steps. (Figure 2.63 shows two of the shortest paths from O to $(3,2)$.)

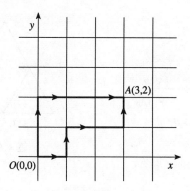

Fig. 2.63

(a) In L_2 the shortest lattice paths from O to $(\bar{x}_1, \bar{x}_2)$ are sequences of $\bar{x}_1 + \bar{x}_2$ steps, any $\bar{x}_1$ of which are taken in the positive direction of the Ox_1 axis, and the remaining x_2 steps are in the positive direction of the Ox_2 axis. Thus the number of shortest paths from O to $(\bar{x}_1, \bar{x}_2)$ is the number of the choices of $\bar{x}_1$ steps from $\bar{x}_1 + \bar{x}_2$ steps, which is equal to the binomial coefficient $C_{\bar{x}_1}^{\bar{x}_1 + \bar{x}_2}$.

(b) In L_3 the shortest lattice paths from O to $(\bar{x}_1, \bar{x}_2, \bar{x}_3)$ are sequences of $\bar{x}_1 + \bar{x}_2 + \bar{x}_3$ steps, in which $\bar{x}_1$ steps are chosen parallel to the positive direction of the Ox_1 axis, a further $\bar{x}_2$ steps are chosen in the positive direction of the Ox_2 axis, and the remaining $\bar{x}_3$ steps have the direction of $+Ox_3$. Hence the number of shortest lattice paths from O to $(\bar{x}_1, \bar{x}_2, \bar{x}_3)$ is equal to the number of the choices of $\bar{x}_1$ elements of one kind, and of $\bar{x}_2$ elements of another kind from a set of $\bar{x}_1 + \bar{x}_2 + \bar{x}_3$ elements. This number is the trinomial coefficient $C_{\bar{x}_1, \bar{x}_2, \bar{x}_3}^{\bar{x}_1 + \bar{x}_2 + \bar{x}_3}$.

(c) In L_n the shortest lattice paths from O to $(\bar{x}_1, \bar{x}_2, \ldots, \bar{x}_n)$ are sequences of $\bar{x}_1 + \bar{x}_2 + \ldots + \bar{x}_n$ steps in which $\bar{x}_i$ steps are chosen in the positive direction of the Ox_i axis for $i = 1, 2, \ldots, n$. Hence the number of shortest lattice paths from O to $(\bar{x}_1, \bar{x}_2, \ldots, \bar{x}_n)$ is equal to the multinomial coefficient $C_{\bar{x}_1, \bar{x}_2, \ldots, \bar{x}_n}^{\bar{x}_1 + \bar{x}_2 + \ldots + \bar{x}_n}$.

Problem 106

See Fig. 2.64.

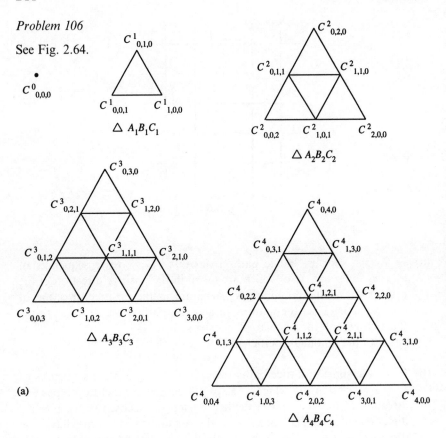

(a)

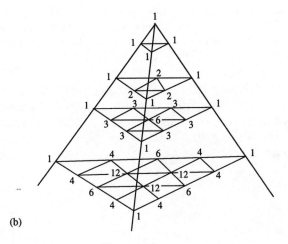

(b)

Fig. 2.64

Problem 107

(a) A shortest lattice path from $O(0,0)$ to $(\bar{x}_1, \bar{x}_2)$, which has no diagonal steps, consists of $\bar{x}_1$ horizontal and $\bar{x}_2$ vertical steps. A diagonal step in a lattice path replaces a horizontal and a vertical step (see Fig. 2.65). Thus a lattice path from O to $(\bar{x}_1, \bar{x}_2)$ with r diagonal steps contains $\bar{x}_1 - r$ horizontal and $\bar{x}_2 - r$ vertical steps.

For a fixed r, the number of lattice paths from O to $(\bar{x}_1, \bar{x}_2)$ with r diagonal steps is equal to the number of choices of r diagonal, $\bar{x}_1 - r$ horizontal, and $\bar{x}_2 - r$ vertical steps from a total of $r + (\bar{x}_1 - r) + (\bar{x}_2 - r) = \bar{x}_1 + \bar{x}_2 - r$ steps. This number is equal to

$$\frac{(\bar{x}_1 + \bar{x}_2 - r)!}{(\bar{x}_1 - r)!(\bar{x}_2 - r)!r!}.$$

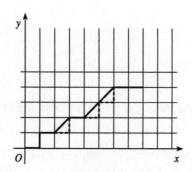

Fig. 2.65

(b) The number r of diagonal steps in a path from O to $(\bar{x}_1, \bar{x}_2)$ can take any value from 0 to the smaller one of the numbers $\bar{x}_1$ and $\bar{x}_2$. Hence the number of all lattice paths admitting diagonal steps is

$$D_{\bar{x}_1, \bar{x}_2} = \sum_{r=0}^{\min(\bar{x}_1, \bar{x}_2)} \frac{(\bar{x}_1 + \bar{x}_2 - r)!}{(\bar{x}_1 - r)!(\bar{x}_2 - r)!r!}.$$

(c) The first few values of $D_{\bar{x}_1, \bar{x}_2}$ are given in Fig. 2.66. This figure shows a table in which the numbers on the kth diagonal d_k (through the points $(k, 0)$ and $(0, k)$) add up to $D_k = \sum_{\bar{x}_1 = 0}^{k} D_{\bar{x}, k - \bar{x}_1}$. The initial values of D_k, $k = 0, 1, 2, 3, 4$, are $1, 2, 5, 12, 29$. This suggests the recurrence formula

$$D_k = 2D_{k-1} + D_{k-2} \quad \text{for} \quad k \geqslant 2. \tag{2.84}$$

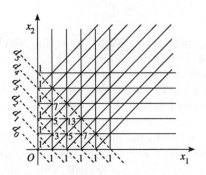

Fig. 2.66

The validity of (2.84) follows from the fact that

$$D_{\bar{x}_1, k - \bar{x}_1} = D_{\bar{x}_1 - 1, (k-1) - (\bar{x}_1 - 1)} + D_{\bar{x}_1, (k-1) - \bar{x}_1} + D_{\bar{x}_1 - 1, (k-2) - (\bar{x}_1 - 1)}$$

$$\text{for} \quad k > 0, \ \bar{x}_1 \neq 0, k$$

and that

$$D_{0,k} = D_{k,0} = 1 \quad \text{for} \quad k \geqslant 0.$$

(2.84) is a special case of the recurrence relation (1.4).

$$f_n(a, b) = a f_{n-1}(a, b) + b f_{n-2}(a, b)$$

for the generalized Fibonacci numbers $f_n(a, b)$. The numbers D_k are generalized Fibonacci numbers $f_n(2, 1)$ with initial values $f_0(2, 1) = 1$ and $f_1(2, 1) = 2$.

Problem 108

(a) In Σ_3 let us label each lattice point (x_1, x_2, x_3) by the number of shortest lattice paths connecting it to $O(0, 0, 0)$, that is by the trinomial coefficient $(x_1 + x_2 + x_3)! / x_1! x_2! x_3!$. Our aim is to find the locus of those lattice points in Σ_3 whose labels are the trinomial coefficients $(\bar{x}_1 + \bar{x}_2 - r)! / (\bar{x}_1 - r)! (\bar{x}_2 - r)! r!$ which appear in the sums for $D_{\bar{x}_1, \bar{x}_2}$ and D_k (see Problem 107).

   We start by establishing the relations

$$\left. \begin{array}{l} x_1 = \bar{x}_1 - r \\ x_2 = \bar{x}_2 - r \\ x_3 = r \end{array} \right\} \tag{2.85}$$

where $\bar{x}_1$ and $\bar{x}_2$ are two given non-negative integers and r is an integer such that $0 \leqslant r \leqslant \min(\bar{x}_1, \bar{x}_2)$.

Put $\bar{x}_1 + \bar{x}_2 = k$. The relations (2.85) lead to the equations

$$x_1 + x_2 + 2x_3 = k \tag{2.86}$$

and

$$x_2 - x_1 = k - 2\bar{x}_1. \tag{2.87}$$

Both equations represent planes in Σ_3. Equation (2.86) is the equation of a plane π_k which intersects the Ox_1, Ox_2 and Ox_3 axes in the points $A_k(k,0,0)$, $B_k(0,k,0)$ and $C_k(0,0,k/2)$ respectively. (2.87) is the equation of a plane $\sigma_{k,i}$ which is perpendicular to $A_k B_k$ and intersects it at the point $E_{k,\bar{x}_1}$. The planes π_k and $\sigma_{k,\bar{x}_1}$ meet along a line $\ell_{k,\bar{x}_1}$ (Fig. 2.67).

The line $\ell_{k,\bar{x}_1}$ carries the lattice points with coordinates (2.85) for all $r = 0, 1, \ldots, \min(\bar{x}, k - \bar{x}_1)$. It is easy to see that these are all the lattice points on $\ell_{k,\bar{x}_1}$. Thus

the locus of lattice points labelled with the trinomial coefficients in the sum for $D_{\bar{x}_1,\bar{x}_2}$ is the set of lattice points on $\ell_{k,\bar{x}_1}$.

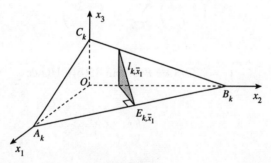

Fig. 2.67

(b) The sum $D_k = \sum_{\bar{x}_1=0}^{k} D_{\bar{x}_1,k-\bar{x}_1}$ contains as summands the trinomial coefficients that are the labels of the lattice points on the lines $\ell_{k,0}, \ell_{k,1}, \ell_{k,2}, \ldots \ell_{k,k}$. The lattice points on all these lines form the set of lattice points of the plane π_k. Hence

the locus of lattice points labelled with the trinomial coefficients of D_k is the set of lattice points on π_k.

Problem 109

(a) The planes perpendicular to the Ox_3 axis and intersecting π_k along lines which contain lattice points, have equations $x_3 = r, r = 0, 1, \ldots, [k/2]$.

Denote the plane with equation $x_3 = r$ by ϑ_r, and its intersection with π_k by $h_{k,r}$ (see Fig. 2.68(a)).

The lattice points on the line $h_{k,r}$ have coordinates $(\bar{x}_1, k - 2r - \bar{x}_1, r)$; here $\bar{x}_1$ is an integer ranging from 0 to $k - 2r$. The trinomial coefficients attached to these points as labels are $(k - r)!/\bar{x}_1!(k - 2r - \bar{x}_1)!r!$. Their sum is

$$S_{k,r} = \sum_{\bar{x}_1 = 0}^{k-2r} \frac{(k-r)!}{\bar{x}_1!(k-2r-\bar{x}_1)!r!}$$

$$= \sum_{\bar{x}_1 = 0}^{k-2r} \frac{(k-r)!}{r!(k-2r)!} \frac{(k-2r)!}{\bar{x}_1!(k-2r-\bar{x}_1)!}$$

$$= \sum_{\bar{x}_1 = 0}^{k-2r} \binom{k-r}{r} \binom{k-2r}{\bar{x}_1}$$

$$= \binom{k-r}{r} \sum_{\bar{x}_1 = 0}^{k-2r} \binom{k-2r}{\bar{x}_1}.$$

$\sum_{\bar{x}_1 = 0}^{k-2r} \binom{k-2r}{\bar{x}_1} = 2^{k-2r}$ (see Problem 23(b)). Hence

$$S_{k,r} = \binom{k-r}{r} \cdot 2^{k-2r}.$$

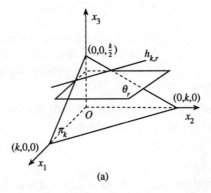

(a)

Fig. 2.68

The union of the sets of lattice points on $h_{k,r}$ for $r = 0, 1, \ldots, [k/2]$ is the set of lattice points on π_k. Hence the sum of the sums $S_{k,r}$ for all r is equal to D_k, defined in Problem 107(c):

$$\sum_{r=0}^{\left[\frac{k}{2}\right]} S_{k,r} = \sum_{r=0}^{\left[\frac{k}{2}\right]} \binom{k-r}{r} 2^{k-2r} = D_k.$$

(b) We wish to find the sum

$$T_{k,i} = \sum_{\bar{x}_3=0}^{\left[\frac{k-i}{2}\right]} \frac{(\bar{x}_1 + i + \bar{x}_3)!}{\bar{x}_1! \, i! \, \bar{x}_3!}.$$

of the trinomial coefficients attached to the lattice points on π_k with a fixed second coordinate $\bar{x}_2 = i$ (Fig. 2.68(b)). Trinomial coefficients have the following important property:

$$\frac{(x_1 + x_2 + x_3)!}{x_1! \, x_2! \, x_3!} = \frac{(x_1 + x_2 + x_3 - 1)!}{(x_1 - 1)! \, x_2! \, x_3!} + \frac{(x_1 + x_2 + x_3 - 1)!}{x_1! \, (x_2 - 1)! \, x_3!}$$

$$+ \frac{(x_1 + x_2 + x_3 - 1)!}{x_1! \, x_2! \, (x_3 - 1)!}.$$

Applying this formula to the trinomial coefficient of a lattice point $P(\bar{x}_1, i, \bar{x}_3)$ on the plane π_k with equation $x_1 + x_2 + 2x_3 = k$ gives

$$\frac{(\bar{x}_1 + i + \bar{x}_3)!}{\bar{x}_1! \, i! \, \bar{x}_3!} = \frac{(\bar{x}_1 + i + \bar{x}_3 - 1)!}{(\bar{x}_1 - 1)! \, i! \, \bar{x}_3!} + \frac{(\bar{x}_1 + i + \bar{x}_3 - 1)!}{\bar{x}_1! \, (i - 1)! \, \bar{x}_3!}$$

$$+ \frac{(\bar{x}_1 + i + \bar{x}_3 - 1)!}{\bar{x}_1! \, i! \, (\bar{x}_3 - 1)!}. \quad (2.88)$$

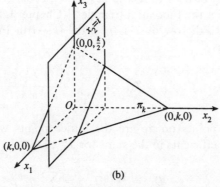

(b)

Fig. 2.68

The points $Q(\bar{x}_1 - 1, i, \bar{x}_3)$ and $R(\bar{x}_1, i - 1, \bar{x}_3)$ lie on the plane π_{k-1} because their coordinates satisfy the equation $x_1 + x_2 + 2x_3 = k - 1$. The point $S(\bar{x}_1, i, \bar{x}_3 - 1)$ is on the plane π_{k-2} because its coordinates satisfy the equation $x_1 + x_2 + 2x_3 = k - 2$. Together with (2.88), this implies that

$$T_{k,i} = T_{k-1,i} + T_{k-1,i-1} + T_{k-2,i} \quad \text{for } k \geqslant 2 \text{ and } 1 \leqslant i \leqslant k. \quad (2.89)$$

For $i = 0$ the trinomial coefficients $(\bar{x}_1 + i + \bar{x}_3)!/\bar{x}_1!\,i!\,\bar{x}_3!$ are equal to the binomial coefficients $(\bar{x}_1 + \bar{x}_3)!/\bar{x}_1!\,\bar{x}_3!$ and relation (2.89) reduces to

$$T_{k,0} = T_{k-1,0} + T_{k-2,0} \quad \text{for } k \geqslant 2. \quad (2.90)$$

It is easy to evaluate the initial values:

$$T_{0,0} = 1 \quad \text{and} \quad T_{1,0} = 1. \quad (2.91)$$

Using (2.89), (2.90), and (2.91) we can construct the following table:

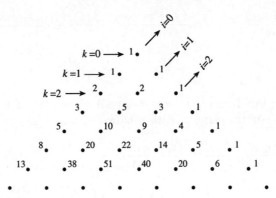

The ith entry of the kth row is $T_{k,i}$.

The table is the Fibonacci triangle T (Chapter I, Section 1.4). The ith entry of the kth row of T is $f_{k-i}^{(i)}$, the $(k - i)$th Fibonacci number of order i. Thus

$$f_{k-i}^{(i)} = T_{k,i} = \sum_{\bar{x}_3 = 0}^{\left[\frac{k-i}{2}\right]} \frac{(\bar{x}_1 + i + \bar{x}_3)!}{\bar{x}_1!\,i!\,\bar{x}_3!}.$$

(c) The lattice points on π_k are the lattice points whose labels are the trinomial coefficients in the sums for $T_{k,i}$, $i = 0, 1, \ldots, k$. Hence

$$D_k = \sum_{i=0}^{k} T_{k,i} = \sum_{i=0}^{k} f_{k-i}^{(i)}.$$

Problem 110

In part (b) of Problem 109 we proved that

$$f_{k-i}^{(i)} = T_{k,i},$$

that is

$$f_{k,i}^{(i)} = \sum_{x_3=0}^{\left[\frac{k-i}{2}\right]} \frac{(x_1+i+x_3)!}{x_1!\,i!\,x_3!}.$$

The points with coordinates (x_1, i, x_3), labelled by the trinomial coefficients in the above sum, belong to the plane π_k with equation $x_1 + x_2 + 2x_3 = k$. Hence $x_1 = k - 2x_3 - i$. Put $x_3 = r$. In this notation

$$f_{k-i}^{(i)} = \sum_{r=0}^{\left[\frac{k-i}{2}\right]} \frac{(k-r)!}{(k-2r-i)!\,i!\,r!} \quad \text{for} \quad i = 0, 1, \ldots, k.$$

For $i = 0$ the above expression yields

$$f_k^{(0)} = \sum_{r=0}^{\left[\frac{k}{2}\right]} \frac{(k-r)!}{(k-2r)!\,r!} = \sum_{r=0}^{\left[\frac{k}{2}\right]} \binom{k-r}{r}.$$

which is Lucas' result for the Fibonacci numbers.

Problem 111

(a) The binomial coefficients $\binom{n}{i}$ satisfy the recurrence relation

$$\binom{n}{i} = \binom{n-1}{i} + \binom{n-1}{i-1}.$$

Hence

$$\rho_{k,n} = \sum_{r=0}^{\left[\frac{k}{2}\right]} \binom{k-r}{r} n^{k-2r} = \sum_{r=0}^{\left[\frac{k}{2}\right]} \left[\binom{k-1-r}{r} + \binom{k-1-r}{r-1}\right] n^{k-2r}$$

$$= \sum_{r=0}^{\left[\frac{k-1}{2}\right]} \binom{k-1-r}{r} n^{k-1-2r} + \sum_{r=0}^{\left[\frac{k-2}{2}\right]} \binom{k-2-r}{r} n^{k-2-2r}$$

Thus $\rho_{k,n}$ satisfies the recurrence relation

$$\rho_{k,n} = n\rho_{k-1,n} + \rho_{k-2,n}.$$

$R_{k,n}$ is a generalized Fibonacci number $f_k(n, 1)$.

(b) The quadratic equation corresponding to the above recurrence relation is

$$x^2 = nx + 1.$$

The solutions of this equation are

$$\alpha = \frac{n + \sqrt{n^2 + 4}}{2} \quad \text{and} \quad \beta = \frac{n - \sqrt{n^2 + 4}}{2}.$$

Thus

$$\rho_{k,n} = A\left(\frac{n + \sqrt{n^2 + 4}}{2}\right)^k + B\left(\frac{n - \sqrt{n^2 + 4}}{2}\right)^k,$$

where A and B can be found with the help of $\rho_{0,n}$ and $\rho_{1,n}$:

$$\left.\begin{array}{r} \rho_{0,n} = 1 = A + B \\ \text{and} \\ \rho_{1,n} = n = A\dfrac{n + \sqrt{n^2 + 4}}{2} + B\dfrac{n - \sqrt{n^2 + 4}}{2} \end{array}\right\}$$

This system of simultaneous equations yields the values

$$A = \frac{n + \sqrt{n^2 + 4}}{2\sqrt{n^2 + 4}} \quad \text{and} \quad B = -\frac{n - \sqrt{n^2 + 4}}{2\sqrt{n^2 + 4}}.$$

This leads to the following generalization of Binet's formula:

$$\rho_{k,n} = \frac{1}{\sqrt{n^2 + 4}}\left[\left(\frac{n + \sqrt{n^2 + 4}}{2}\right)^{k+1} - \left(\frac{n - \sqrt{n^2 + 4}}{2}\right)^{k+1}\right].$$

Problem 112

(a) Let s be an arbitrary sequence of n bets each of which results in a gain of £1 or in a loss of £1. In the lattice path ℓ representing s, each gain of £1 corresponds to a horizontal step ($\rightarrow$) and each loss of £1 to a vertical step ($\uparrow$). Since ℓ starts at $O(0,0)$, this implies that it ends on the line segment AB with endpoints $A(0, n)$ and $B(n, 0)$.

Any lattice point $P(\bar{x}, \bar{y})$ of ℓ represents the gambler's fortune after the first $\bar{x} + \bar{y}$ bets; at this stage the gambler has (or lacks) $p + \bar{x} - \bar{y}$ pounds. Bankruptcy occurs if and only if, after a certain number of bets, $p + \bar{x} - \bar{y}$ becomes 0. This will happen if and only if ℓ meets the line ℓ_1 with equation $y = x + p$. A lattice path ℓ can meet ℓ_1 only at a lattice point of the segment PQ with endpoints $P(0, p)$ and $Q \in AB$ (see Fig. 2.69). Thus

> a lattice path ℓ represents a ruinous sequence if and only if it meets PQ at least once.

A lattice path meeting PQ at a point $(\bar{x}, p + \bar{x})$ cannot end below the line ℓ_2 with equation $y = p$. Hence all ruinous lattice paths have their endpoints on the line segment AC with $C = \ell_2 \cap AB$.

(b) We start with a number of observations.

(1) The point Q has coordinates $((n - p)/2, (n + p)/2)$, which are not necessarily integers. Thus Q is not necessarily a lattice point. If Q is a lattice point then the number of lattice paths ending at Q is
$$\binom{n}{(n-p)/2}.$$

(2) The set of those lattice paths which meet PQ at points different from Q can be divided into two disjoint subsets: S_1, the set of lattice paths with endpoints on AQ, and S_2, the set of lattice paths with endpoints on QC.

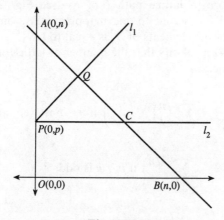

Fig. 2.69

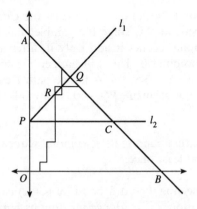

Fig. 2.70

(3) The number $|S_1|$ of lattice paths in S_1 is given by

$$|S_1| = \binom{n}{0} + \binom{n}{1} + \ldots + \binom{n}{\frac{n-p}{2} - 1} \quad \text{if } Q \text{ is a lattice point, and}$$

$$|S_1| = \binom{n}{0} + \binom{n}{1} + \ldots + \binom{n}{\frac{n-p-1}{2}} \quad \text{if } Q \text{ is not a lattice point.}$$

(4) Each element $\ell \in S_1$ can be mapped onto a unique element $\hat{\ell} \in S_2$ by the following transformation.

Let R be the first point where ℓ meets PQ on the way from O to its endpoint. Divide ℓ into two parts: ℓ_{I} from O to R and ℓ_{II} from R to AC. Construct the reflection $\hat{\ell}_{\mathrm{II}}$ of ℓ_{II} in the line PQ. The parts ℓ_{I} and $\hat{\ell}_{\mathrm{II}}$ together form a lattice path $\hat{\ell}$ of S_2 (see Fig. 2.70). The transformation $\ell \to \hat{\ell}$ is a one-to-one mapping of S_1 onto S_2. Thus the number $|S_2|$ of the elements of S_2 is equal to $|S_1|$.

From (1)–(4) it follows that the number of lattice paths representing ruinous sequences is

$$2 \cdot \sum_{i=0}^{\frac{n-p}{2} - 1} \binom{n}{i} + \binom{n}{\frac{n-p}{2}} \quad \text{if } n - p \text{ is even, and}$$

$$2 \cdot \sum_{i=0}^{\frac{n-p-1}{2} - 1} \binom{n}{i} \quad \text{if } n - p \text{ is odd.}$$

Chapter IX

Problem 113

(a) Let n be an arbitrary integer. $\oplus$ is a binary operation on $\mathbb{Z}_n = \{0, 1, \ldots, n-1\}$ because $i \oplus j \in \mathbb{Z}_n$ for any $i, j \in \mathbb{Z}_n$. It is easy to verify that $\mathbb{Z}_n$ ($\oplus$) satisfies all group axioms:

G_1: $(i \oplus j) \oplus k = (i \oplus (j \oplus k)$ for all $i, j, k \in \mathbb{Z}_n$.

G_2: $0 \oplus i = i \oplus 0 = i$ for all $i \in \mathbb{Z}_n$. Thus 0 is the identity element of $\mathbb{Z}_n$ ($\oplus$).

G_3: $i \oplus (n-i) = (n-i) \oplus i = 0$ for $i = 1, 2, \ldots, n-1$ and $0 \oplus 0 = 0$, that is, every element of $\mathbb{Z}_n$ ($\oplus$) has an inverse.

Hence $\mathbb{Z}_n$ ($\oplus$) is a group for any natural number n.

(b) Two cases must be distinguished:

(i) If n is not a prime number, then $\mathbb{Z}_n^* = \{1, 2, \ldots, n-1\}$ contains two elements n' and n'', different from 1, such that $n' \cdot n'' = n$. Thus $n' \otimes n'' = 0 \notin \mathbb{Z}_n^*$. In other words, $\otimes$ is not a binary operation on $\mathbb{Z}_n^*$. It follows that $\mathbb{Z}_n^*$ ($\otimes$) is not a group for any composite number n.

(ii) If n is a prime number p, then $i \cdot j \neq p$ for all $i, j \in \mathbb{Z}_p^*$. Hence $i \otimes j \in \mathbb{Z}_p^*$ for all $i, j \in \mathbb{Z}_p^*$. $\otimes$ is a binary operation on $\mathbb{Z}_n^*$. It is left to the reader to verify the group axioms G_1–G_3 for $\mathbb{Z}_p^*$.

From (i) and (ii) it follows that $\mathbb{Z}_n^*$ ($\otimes$) is a group if and only if n is a prime.

Problem 114

Figure 2.71 shows the so-called operation table for $S = \{f_1(x), f_2(x), f_3(x)\}$ with respect to $\circ$. The entry in the ith row and in the jth column of the table is $f_i(x) \circ f_j(x)$, that is $f_i(f_j(x))$. The table shows that:

(1) $f_i(x) \circ f_j(x) \in S$ for all $f_i(x), f_j(x) \in S$. In other words, $\circ$ is a binary operation on S.

(2) $f_1(x)$ is the identity element of $S(\circ)$

(3) $f_1(x)$, $f_3(x)$, and $f_2(x)$ are the inverse elements of $f_1(x)$, $f_2(x)$, and $f_3(x)$ respectively in $S(\circ)$.

$\circ$	$f_1(x)$	$f_2(x)$	$f_3(x)$
$f_1(x)$	$f_1(x)$	$f_2(x)$	$f_3(x)$
$f_2(x)$	$f_2(x)$	$f_3(x)$	$f_1(x)$
$f_3(x)$	$f_3(x)$	$f_1(x)$	$f_2(x)$

Fig. 2.71

It would be tedious to check the associativity of $\circ$ for all triples $f_i(x)$, $f_j(x)$, $f_k(x)$, $i,j,k \in \{1,2,3\}$. Instead we shall rely on the well known fact that for composition of mappings (provided the composed mappings are defined) the associative law holds.

Thus S forms a group under $\circ$.

Problem 115

Denote by $\mathscr{F}$ the set of all functions $f_{a,b,c,d}(x)$.

First we have to verify that $f_{a,b,c,d}(x) \circ f_{r,s,t,u}(x) \in \mathscr{F}$ for any $f_{a,b,c,d}(x), f_{r,s,t,u}(x) \in \mathscr{F}$.

In the general case the definition of $\circ$ implies that

$$f_{a,b,c,d}(x) \circ f_{r,s,t,u}(x) = \frac{a(rx+s)/(tx+u)+b}{c(rx+s)/(tx+u)+d} = \frac{(ar+bt)x+as+bu}{(cr+dt)x+cs+du}$$

$$= \frac{kx+\ell}{mx+n} = f_{k,\ell,m,n}(x),$$

where $k = ar + bt$, $\ell = as + bu$, $m = cr + dt$, and $n = cs + du$.

From $ad - bc = 1$ and $ru - st = 1$ it follows that k, ℓ, m, n are integers satisfying the condition

$$kn - m\ell = (ar+bt)(sc+du) - (as+bu)(cr+dt) = (ru-st)(ad-bc) = 1.$$

If $m = 0$ then $n \neq 0$ (otherwise $kn - m\ell$ would be equal to 0, contrary to $kn - m\ell = 1$).

By analysing all possible cases when $m = 0$ or $m \neq 0$ we find that

(a) if $m \neq 0$ then

$$f_{k,\ell,m,n}(x) = \begin{cases} \dfrac{kx+\ell}{mx+n} & \text{if } x \neq -\dfrac{n}{m} \text{ and } x \neq \infty \\[2mm] \dfrac{k}{m} & \text{if } x = \infty \\[2mm] \infty & \text{if } x = -\dfrac{n}{m}, \end{cases}$$

(b) if $m = 0$ then

$$f_{k,\ell,m,n}(x) = \begin{cases} \dfrac{kx + \ell}{n} & \text{if } x \neq \infty \\ \infty & \text{if } x = \infty. \end{cases}$$

Hence $f_{k,\ell,m,n}(x) \in \mathscr{F}$; the composition of functions is a binary operation on $\mathscr{F}$.

The function $f_{1,0,0,1}(x) = x$ is the identity of $\mathscr{F}$ ($\circ$).

The inverse of $f_{a,b,c,d}(x) \in \mathscr{F}(x)$ with respect to $\circ$ if $f_{d,-b,-c,a}(x)$.

The associativity of $\circ$ is a consequence of the fact that addition and multiplication of integers are associative and commutative operations. (An operation $*$ on a set S is commutative if $a * b = b * a$ for all $a, b \in S$.)

Thus $\mathscr{F}$ ($\circ$) is a group.

Problem 116

(a) Denote the set of all permutations on S (that is, the set of all one-to-one mappings of S onto itself) by σ_n, where n is the number of elements in S.

For any $\pi, \pi' \in \sigma_n$ the mapping $\pi \circ \pi'$, defined by $\pi \circ \pi'(x) = \pi(\pi'(x))$ for all $x \in S$, is a one-to-one mapping of S itself. Hence $\pi \circ \pi' \in \sigma_n$; in other words, the composition of permutations is a binary operation on σ_n.

$[\pi \circ (\pi' \circ \pi'')](x) = [(\pi \circ \pi') \circ \pi''](x)$ for all $x \in S$ and all $\pi, \pi', \pi'' \in \sigma_n$.

Hence the operation $\circ$ is associative.

The permutation which leaves every element x of S unchanged is the identity element of σ_n.

The inverse of an arbitrary permutation $\pi \in S$ is the permutation $\pi^{-1}(x) \in \sigma_n$ such that $\pi^{-1}(\pi(x)) = x$ for all $x \in S$.

The above facts imply that $\sigma_n(\circ)$ is a group.

(b) Let $S = \{x_1, x_2, \ldots, x_n\}$ be a set of n elements. An arbitrary permutation π of S can be constructed as follows.

Choose an element $x_i \in S$ and put $\pi(x_1) = x_i$. The element x_i can be chosen in n different ways. For a fixed i there are $n - 1$ choices of $x_j \in \{x_1, x_2, \ldots, x_{i-1}, x_{i+1}, \ldots, x_n\}$ such that $\pi(x_2) = x_j$. For fixed i and j there are $n - 2$ choices of $x_k \in S$: $x_k \neq x_i, x_j$ such that $\pi(x_3) = x_k$, and so on.

Thus there are altogether $n(n - 1)(n - 2)\ldots 3.2.1 = n!$ permutations of S onto itself.

Problem 117

(a) We consider the fraction

$$F_\pi = \frac{(a_1 - a_2)(a_1 - a_3)\ldots(a_1 - a_n)}{[\pi(a_1) - \pi(a_2)][\pi(a_1) - \pi(a_3)]\ldots[\pi(a_1) - \pi(a_n)]}.$$

$$\frac{(a_2 - a_3)(a_2 - a_4)\ldots(a_2 - a_n)}{[\pi(a_2) - \pi(a_3)][\pi(a_2) - \pi(a_4)]\ldots[\pi(a_2) - \pi(a_n)]}\ldots\frac{a_{n-1} - a_n}{\pi(a_{n-1}) - \pi(a_n)}.$$

Every $a_i - a_j$ with $i < j$ occurs in the numerator as a factor. The denominator also contains the factor $a_i - a_j$ if the order of a_i and a_j in the bottom row is not inverted. Otherwise the denominator contains $a_j - a_i = -(a_i - a_j)$. But then the fraction reduces to $(-1)^v$, where v is the number of inversions of π. Thus $F_\pi = \text{sign } \pi$.

(b) Let π and π' be two (not necessarily distinct) permutations of S. According to (a),

$$\text{sign } \pi = \underset{i<j}{\Pi}\frac{a_i - a_j}{\pi(a_i) - \pi(a_j)}, \quad \text{sign } \pi' = \underset{i<j}{\Pi}\frac{a_i - a_j}{\pi'(a_i) - \pi'(a_j)}$$

and

$$\text{sign } \pi \circ \pi' = \underset{i<j}{\Pi}\frac{a_i - a_j}{\pi(\pi'(a_i)) - \pi(\pi'(a_j))}.$$

Put $b_i = \pi'(a_i)$, $i = 1,\ldots,n$. Then the product, representing sign $(\pi \circ \pi')$ can be rewritten as

$$\underset{i<j}{\Pi}\frac{a_i - a_j}{\pi'(a_i) - \pi'(a_j)} \cdot \frac{\pi'(a_i) - \pi'(a_j)}{\pi(\pi'(a_i)) - \pi(\pi'(a_j))}$$

$$= \underset{i<j}{\Pi}\frac{a_i - a_j}{\pi'(a_i) - \pi'(a_j)} \cdot \frac{b_i - b_j}{\pi(b_i) - \pi(b_j)}$$

$$= \text{sign } \pi' \cdot \text{sign } \pi.$$

Problem 118

(a) Let A_n be the subset of the even permutations in σ_n. The identity permutation ε belongs to σ_n because sign $\varepsilon = 1$. From sign $(\pi \circ \pi') = \text{sign } \pi \cdot \text{sign } \pi'$ it follows that the product of two permutations is even if and only if both permutations have the same sign. Thus

the product of two even permutations is even, and so is the inverse of an even permutation.

Hence A_n is a subgroup of σ_n.

(b) Let $S = \{s_1, s_2, \ldots, s_n\}$ be a set of $n > 1$ elements. Denote by π^* the permutation which interchanges s_1 and s_2, but fixes s_i for $i = 3, \ldots, n$. The permutation π^* is odd, moreover $\pi^* \circ \pi^* = \varepsilon$.

Denote the set of odd permutations in σ_n by $\mathcal{O}_n$. It is easy to verify that

(i) $\pi^* \circ \pi \in \mathcal{O}_n$ for all $\pi \in A_n$,

(ii)   for any $\tilde{\pi} \in \mathcal{O}_n$ there is an element $\hat{\pi} \in A_n$ such that $\tilde{\pi} = \pi^* \circ \hat{\pi}$
       $(\hat{\pi} = \pi^* \circ \tilde{\pi})$

(iii) $\pi^* \circ \pi = \pi^* \circ \pi'$ implies that $\pi = \pi'$ for all $\pi, \pi' \in A_n$.

In view of (i)–(iii) the mapping $\pi \to \pi^* \circ \pi$ for all $\pi \in A_n$ establishes a one-to-one correspondence between the elements of A_n and $\mathcal{O}_n$. Thus A_n and $\mathcal{O}_n$ contain $\frac{1}{2}n!$ elements each.

Problem 119

(a) The permutations of $S = \{1, 2, 3\}$ are

$$\pi_1 = \begin{pmatrix} 1 & 2 & 3 \\ 1 & 2 & 3 \end{pmatrix}, \quad \pi_2 = \begin{pmatrix} 1 & 2 & 3 \\ 2 & 3 & 1 \end{pmatrix}, \quad \pi_3 = \begin{pmatrix} 1 & 2 & 3 \\ 3 & 1 & 2 \end{pmatrix},$$

$$\pi_4 = \begin{pmatrix} 1 & 2 & 3 \\ 1 & 3 & 2 \end{pmatrix}, \quad \pi_5 = \begin{pmatrix} 1 & 2 & 3 \\ 3 & 2 & 1 \end{pmatrix}, \quad \text{and} \quad \pi_6 = \begin{pmatrix} 1 & 2 & 3 \\ 2 & 1 & 3 \end{pmatrix}.$$

They form the symmetric group σ_3. The alternating group A_3 of this group consists of the permutations π_1, π_2, and π_3.

Since $\pi_3 = \pi_2 \circ \pi_2$ and $\pi_1 = \pi_2 \circ \pi_2 \circ \pi_2$, the group A_3 is generated by π_2. This implies that

$$\pi_2, \pi_2 \circ \pi_2, \quad \pi_2 \circ \pi_2 \circ \pi_2, \quad \pi_4 \circ \pi_2 (= \pi_5), \quad \pi_4 \circ \pi_2 \circ \pi_2 (= \pi_6),$$
$$\text{and} \quad \pi_4 \circ \pi_2 \circ \pi_2 \circ \pi_2 (= \pi_4) \text{ are the six elements of } \sigma_3.$$

Thus π_2 and π_4 form a set of generators for σ_3.

Another set of generators, which consists of two odd permutations, is $\{\pi_4, \pi_5\}$. In this case

$$\pi_2 = \pi_4 \circ \pi_5, \quad \pi_1 = \pi_4 \circ \pi_5 \circ \pi_4 \circ \pi_5 \circ \pi_4 \circ \pi_5,$$
$$\pi_3 = \pi_4 \circ \pi_5 \circ \pi_4 \circ \pi_5, \quad \text{and} \quad \pi_6 = \pi_4 \circ \pi_4 \circ \pi_5 \circ \pi_4 \circ \pi_5.$$

σ_3 has a well known geometric interpretation: it is the symmetry group of an equilateral triangle. In Fig. 2.72 π_1, π_2, and π_3 correspond to the rotations of the triangle $A_1 A_2 A_3$ about its centre O through $0°$, $120°$, and $240°$ respectively. π_4, π_5, and π_6 correspond to the reflections of the triangle in the lines $A_1 O$, $A_2 O$, and $A_3 O$.

(b) Any element $i \in \mathbb{Z}_n$ can be written as the 'sum' $\underbrace{1 \oplus 1 \oplus 1 \oplus \ldots \oplus 1}_{i \text{ times}}$.

Thus $\mathbb{Z}_n$ can be generated by a single element, 1.

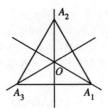

Fig. 2.72

Problem 120

The functions $g_1(x) = x$ and $g_2(x) = -1/x$ form a group under the operation of composition of functions. Rewrite the functional equation

$$xf(x) + 2f\left(-\frac{1}{x}\right) = 3 \tag{2.92}$$

as

$$g_1(x)f(g_1(x)) + 2f(g_2(x)) = 3,$$

and substitute $g_2(x)$ for x. This yields

$$g_1(g_2(x))f(g_1(g_2(x))) + 2f(g_2(g_2(x))) = 3,$$

that is

$$-\frac{1}{x}f\left(-\frac{1}{x}\right) + 2f(x) = 3. \tag{2.93}$$

By eliminating $f\left(-\dfrac{1}{x}\right)$ from the system of equations (2.92) and (2.93) we find that

$$f(x) = \frac{6x + 3}{5x}. \tag{2.94}$$

hence

$$f\left(-\frac{1}{x}\right) = \frac{6 - 3x}{5}. \tag{2.95}$$

By substituting (2.94) and (2.95) in (2.92) we can check that

$$x\frac{6x + 3}{5x} + 2\frac{6 - 3x}{5} = 3.$$

Hence the solution of the functional equation (2.92) is $f(x) = (6x + 3)/5x$ for all real numbers $x \neq 0$.

Problem 121

The functions $g_1(x) = x$ and $g_2(x) = -x$ form a group under the operation of composition of functions.

By substituting $g_2(x) = -x$ for x in

$$af(x^n) + f(-x^n) = bx \tag{2.96}$$

we find that

$$af((-x)^n) + f(-(-x)^n) = b(-x),$$

or, since n is an odd integer,

$$af(-x^n) + f(x^n) = -bx. \tag{2.97}$$

(2.96) and (2.97) imply that

$$f(x^n) = \frac{bx}{a-1};$$

hence

$$f(x) = \frac{b\sqrt[n]{x}}{a-1}. \tag{2.98}$$

$\sqrt[n]{x}$ is defined for all real numbers x, since n is odd; $a \neq 1$.

Thus $f(x)$ is defined for all real numbers x. Moreover,

$$af(x^n) + f(-x^n) = \frac{ab}{a-1}x - \frac{b}{a-1}x = \frac{a-1}{a-1}bx = bx.$$

This proves that (2.98) is the solution of our functional equation.

Problem 122

(a) The addition of two arbitrary vectors $\begin{pmatrix} x_1 \\ y_1 \end{pmatrix}$ and $\begin{pmatrix} x_2 \\ y_2 \end{pmatrix}$ with real entries x_1, y_1, x_2 and y_2 is defined by

$$\begin{pmatrix} x_1 \\ y_1 \end{pmatrix} + \begin{pmatrix} x_2 \\ y_2 \end{pmatrix} = \begin{pmatrix} x_1 + x_2 \\ y_1 + y_2 \end{pmatrix}.$$

Since the sums $x_1 + x_2$ and $y_1 + y_2$ are real numbers, this implies that $V = \left\{ \begin{pmatrix} x \\ y \end{pmatrix}; x, y \text{ real numbers} \right\}$ is closed under vector addition.

The identity of $V(+)$ is the 'zero vector' $\begin{pmatrix} 0 \\ 0 \end{pmatrix}$, and the inverse of $\begin{pmatrix} x \\ y \end{pmatrix} \in V$ is $\begin{pmatrix} -x \\ -y \end{pmatrix}$.

Vector addition is associative because addition of real numbers is associative. Indeed,

$$\left[\begin{pmatrix} x_1 \\ y_1 \end{pmatrix} + \begin{pmatrix} x_2 \\ y_2 \end{pmatrix}\right] + \begin{pmatrix} x_3 \\ y_3 \end{pmatrix} = \begin{pmatrix} x_1 + x_2 \\ y_1 + y_2 \end{pmatrix} + \begin{pmatrix} x_3 \\ y_3 \end{pmatrix} = \begin{pmatrix} (x_1 + x_2) + x_3 \\ (y_1 + y_2) + y_3 \end{pmatrix}$$

$$= \begin{pmatrix} x_1 + (x_2 + x_3) \\ y_1 + (y_2 + y_3) \end{pmatrix} = \begin{pmatrix} x_1 \\ y_1 \end{pmatrix} + \begin{pmatrix} x_2 + x_3 \\ y_1 + y_2 + y_3 \end{pmatrix}$$

$$= \begin{pmatrix} x_1 \\ y_1 \end{pmatrix} + \left[\begin{pmatrix} x_2 \\ y_2 \end{pmatrix} + \begin{pmatrix} x_3 \\ y_3 \end{pmatrix}\right].$$

Thus $V(+)$ is a group.

(b) For arbitrary integers m_1, m_2 the numbers $m_1 + m_2$ and $-m_1$ are integers. Hence the set H of vectors $\begin{pmatrix} m \\ n \end{pmatrix}$, where m and n are arbitrary integers, forms a subgroups of $V(+)$.

Problem 123 ·

(a) An arbitrary vector $\begin{pmatrix} m \\ n \end{pmatrix}$ of $H(+)$ can be expressed as

$$\begin{pmatrix} m \\ n \end{pmatrix} = \begin{pmatrix} m \\ 0 \end{pmatrix} + \begin{pmatrix} 0 \\ n \end{pmatrix} = \underbrace{\begin{pmatrix} 1 \\ 0 \end{pmatrix} + \begin{pmatrix} 1 \\ 0 \end{pmatrix} + \ldots + \begin{pmatrix} 1 \\ 0 \end{pmatrix}}_{m \text{ times}} + \underbrace{\begin{pmatrix} 0 \\ 1 \end{pmatrix} + \begin{pmatrix} 0 \\ 1 \end{pmatrix} + \ldots + \begin{pmatrix} 0 \\ 1 \end{pmatrix}}_{n \text{ times}}.$$

Thus the vectors $OI = \begin{pmatrix} 1 \\ 0 \end{pmatrix}$ and $OJ = \begin{pmatrix} 0 \\ 1 \end{pmatrix}$ generate $H(+)$.

(b) First we shall prove the following.

(i) Two vectors $\begin{pmatrix} a \\ b \end{pmatrix}$ and $\begin{pmatrix} c \\ d \end{pmatrix}$ of $H(+)$ generate $H(+)$ if and only if there exist integers $p, q, r,$ and s such that

$$\left.\begin{aligned} \begin{pmatrix} 1 \\ 0 \end{pmatrix} &= p\begin{pmatrix} a \\ b \end{pmatrix} + q\begin{pmatrix} c \\ d \end{pmatrix} \\ \begin{pmatrix} 0 \\ 1 \end{pmatrix} &= r\begin{pmatrix} a \\ b \end{pmatrix} + s\begin{pmatrix} c \\ d \end{pmatrix}. \end{aligned}\right\} \qquad (2.99)$$

and

For proof note that if (2.99) holds, then an arbitrary vector $\begin{pmatrix} m \\ n \end{pmatrix} \in H(+)$ can be written as

$$\binom{m}{n} = m\binom{1}{0} + n\binom{0}{1} = m\left[p\binom{a}{b} + q\binom{c}{d}\right] + n\left[r\binom{a}{b} + s\binom{c}{d}\right]$$

$$= (mp + nr)\binom{a}{b} + (mq + ns)\binom{c}{d}$$

$$= \underbrace{\binom{a}{b} + \binom{a}{b} + \ldots + \binom{a}{b}}_{mp+nr \text{ times}} + \underbrace{\binom{c}{d} + \binom{c}{d} + \ldots + \binom{c}{d}}_{mq+ns \text{ times}}.$$

Thus $\binom{a}{b}$ and $\binom{c}{d}$ generate $H(+)$.

Conversely, if $\binom{a}{b}$ and $\binom{c}{d}$ generate $H(+)$, then any vector $\binom{m}{n} \in H(+)$ is expressible as $u\binom{a}{b} + v\binom{c}{d}$ for appropriate integers u and v. Hence, in particular, relations (2.99) are valid.

The next step is to show the following:

(ii) Let $\binom{a}{b}$ and $\binom{c}{d}$ be two vectors in $H(+)$. The necessary and sufficient condition for the existence of integers p, q, r, and s such that relations (2.99) are satisfied is the equality

$$|ad - bc| = 1.$$

(ii) can be proved as follows.

Conditions (2.99) yield two systems of simultaneous equations:

$$\text{(I)} \quad \begin{array}{l} ap + cq = 1 \\ bp + dq = 0 \end{array} \quad \text{and} \quad \text{(II)} \quad \begin{array}{l} ar + cs = 0 \\ br + ds = 1. \end{array}$$

Both systems have the same determinant $ad - bc$. It is easy to verify that both systems have rational solutions if and only if $ad - bc \neq 0$. The solutions are

$$p = \frac{d}{ad - bc}, \quad q = \frac{-b}{ad - bc}, \quad r = \frac{-c}{ad - bc}, \quad \text{and} \quad s = \frac{a}{ad - bc}.$$

If $|ad - bc| = 1$, then p, q, r, and s are integers.

Conversely, suppose that p, q, r and s are integers. In that case a, b, c, and d must all be divisible by $ad - bc$, and therefore

$$\frac{a}{ad - bc} \cdot \frac{d}{ad - bc} - \frac{b}{ad - bc} \cdot \frac{c}{ad - bc} = \frac{ad - bc}{(ad - bc)^2}$$

must be an integer. This can happen only if $ad - bc = \pm 1$, that is if $|ad - bc| = 1$.

(i) and (ii) imply that

$\binom{a}{b}$ and $\binom{c}{d}$ of $H(+)$ generate $H(+)$ if and only if $|ad - bc| = 1$.

Problem 124

We shall present two solution methods.

The first method consists in circumscribing the smallest possible rectangle $\overline{R}$ about the parallelogram $OPRQ$, with sides parallel to the coordinate axes. The area of $OPRQ$ is the difference of the area of $\overline{R}$ and of the areas of triangles in $\overline{R}$, outside $OPRQ$ (see Fig. 2.73). It is easy to conclude that

$$\text{Area } OPRQ = |ab - cd|.$$

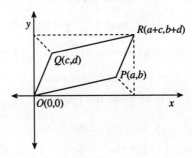

Fig. 2.73

The second method uses vectors. In a cartesian coordinate system with perpendicular axes $O'x$, $O'y$ and $O'z$ consider the parallelogram with vertices $O'(0,0,0)$, $P'(a,b,0)$, $R'(a+c,b+d,0)$ and $Q'(c,d,0)$. Denote the vectors

$$\begin{pmatrix} 1 \\ 0 \\ 0 \end{pmatrix}, \begin{pmatrix} 0 \\ 1 \\ 0 \end{pmatrix} \quad \text{and} \quad \begin{pmatrix} 0 \\ 0 \\ 1 \end{pmatrix}$$

by i, j, k respectively (Fig. 2.74). It is well known that the numerical value of the area of $O'P'Q'R'$ is equal to $|O'P' \times O'Q'|$, where $O'P' \times O'Q'$ is the vector product of

$$\begin{pmatrix} a \\ b \\ 0 \end{pmatrix} \quad \text{and} \quad \begin{pmatrix} c \\ d \\ 0 \end{pmatrix}.$$

Since

$$O'P' \times O'Q' = \begin{vmatrix} i & j & k \\ a & b & 0 \\ c & d & 0 \end{vmatrix} = (ad - bc)k,$$

it follows that

$$\text{Area } O'P'Q'R' = |(ad - bc)k| = |ad - bc|$$

for all real numbers $a, b, c,$ and d.

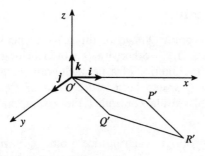

Fig. 2.74

Problem 125

(a) Consider first the special case when P is a parallelogram $P_{0,0}$ with one
    vertex $A_{0,0} = O(0,0)$. The remaining three vertices of $P_{0,0}$ are lattice
    points $A_{1,0}(a,b)$, $A_{0,1}(c,d)$ and $A_{1,1}(a+c,b+d)$ (Fig. 2.75). According
    to our assumption, there are no other lattice points on the perimeter
    or in the interior of $P_{0,0}$. For arbitrary integers i, j denote by $P_{i,j}$ the
    parallelogram whose vertices are the lattice points $A_{i,j}(ia+jc, ib+jd)$,
    $A_{i+1,j}((i+1)a+jc, (i+1)b+jd)$, $A_{i+1,j+1}((i+1)a+(j+1)c, (i+1)b+$
    $(j+1)d)$, and $A_{i,j+1}(ia+(j+1)c, ib+(j+1)d)$. The parallelogram $P_{i,j}$
    is the image of $P_{0,0}$ under the translation $\tau_{i,j}$ with translation vector
    $\begin{pmatrix} ia+jc \\ ib+jd \end{pmatrix}$. The parallelograms $P_{i,j}$ for $i,j \in \{0, \pm 1, \pm 2, \ldots\}$ tessellate
    the plane.

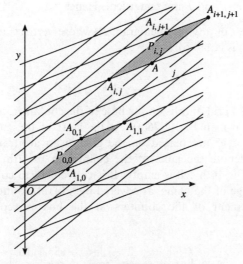

Fig. 2.75

This implies that:

(i) Each lattice point $Q(m, n)$ of the plane coincides with one of the lattice points $A_{i,j}$ —otherwise Q would belong to the inside of a side, or to the interior of a parallelogram $P_{i',j'}$. Hence the image $\tau_{i',-j'}(Q)$ of Q would be a lattice point of $P_{0,0}$ different from its vertices. This would contradict the assumption about the lattice points of $P_{0,0}$.

In view of (i), each lattice point $Q(m, n)$ has coordinates $m = ia + jc$, $n = ib + jd$ for some integers i, j. This implies that each vector $\binom{m}{n}$ of $H(+)$ is expressible as $i\binom{a}{b} + j\binom{c}{d}$. In other words,

(ii) $\binom{a}{b}$ and $\binom{c}{d}$ generate $H(+)$.

(ii) implies that $|ad - bc| = 1$ (see Problem 122(b)). Since $|ad - bc|$ is the area of $P_{0,0}$ (this was proved in Problem 123), the area of $P_{0,0}$ is 1.

(b) Let P be a parallelogram such that its only lattice points are its vertices $A(p, q)$, $B(p + a, q + b)$, $C(p + a + c, q + b + d)$, $D(p + c, q + d)$. The image of P under the translation $\tau_{p,-q}$ with translation vector $\binom{-p}{-q}$ is a parallelogram $P_{0,0}$ with a vertex $O(0, 0)$ and with no lattice points except its vertices. According to (a), the area of $P_{0,0}$ is 1. The parallelograms $P_{0,0}$ and P are congruent. Hence

the area of any parallelogram P, whose vertices are its only lattice points, is equal to 1.

Problem 126

We draw a 4×4 grid of cells on the base of the container, and label the cells 1 to 16, as shown in Fig. 2.76(a). At the start of the game the counters labelled $1, 2, \ldots, 15$ are placed on the grid in an arbitrary order. In this arrangement denote the label of the counter in the ith cell by $\pi_0(i)$ for $i = 1, \ldots, 15$. The 16th cell is empty. It is convenient to consider the empty cell, at any stage of the game, as covered by a blank counter '$\bigcirc$'. Thus the initial arrangement of the counters in the box is represented by the permutation

$$\pi_0 = \begin{pmatrix} 1 & 2 & & 15 & 16 \\ \pi_0(1) & \pi_0(2) & \cdots & \pi_0(15) & \bigcirc \end{pmatrix}.$$

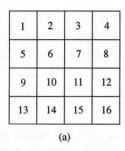

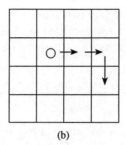

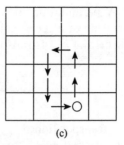

(a)   (b)   (c)

Fig. 2.76

As the counters $1, \ldots, 15$ are shifted around, the blank counter $\bigcirc$ appears in different cells; it is 'moving about' in the box. By studying its moves we can investigate the parity of the permutations representing the arrangements of the counters during the game.

An arbitrary route of $\bigcirc$ may consist of two types of paths:

(I)   paths leading through cells with consecutive labels (in increasing or in decreasing order) (Fig. 2.76(b)), and

(II) closed cyclic paths (Fig. 2.76(c)).

In a path of type (I) no move of $\bigcirc$ changes the order in which the remaining counters follow one another. Hence

(i) path of type (I) do not change the parity of the permutation corresponding to the previous arrangement of the counters.

Now consider any path P of type II. It consists of a circuit of cells, labelled, say, $i_1, i_2, \ldots, i_k$. The format of the grid implies that the number k of cells in the circuit is even. Suppose that at a certain stage of the game the blank counter is in the i_kth cell, and that at the next step the following changes take place: the counter in the cell i_1 moves to the i_kth cell, the counters in the cells i_j move to the cells i_{j-1} for $j = 2, \ldots, k$, and $\bigcirc$ moves to cell i_{k-1}. The corresponding permutations, before and after this step, are

$$\pi = \begin{pmatrix} i_1 & i_2 & i_3 & \ldots & i_{k-1} & i_k & \ldots & p & \ldots & q & \ldots \\ \pi(i_1) & \pi(i_2) & \pi(i_3) & \ldots & \pi(i_{k-1}) & \bigcirc & \ldots & \pi(p) & \ldots & \pi(q) & \ldots \end{pmatrix}$$

and

$$\pi' = \begin{pmatrix} i_1 & i_2 & i_3 & \ldots & i_{k-1} & i_k & \ldots & p & \ldots & q & \ldots \\ \pi(i_2) & \pi(i_3) & \pi(i_4) & \ldots & \bigcirc & \pi(i_1) & \ldots & \pi(p) & \ldots & \pi(q) & \ldots \end{pmatrix}$$

Denote by $\vartheta_{i_j, i_{j+1}}$ the permutation of $S = \{1, 2, \ldots, 16\}$ which interchanges the elements i_j and i_{j+1}, and leaves the remaining elements of S fixed.

A permutation interchanging two elements, and fixing the remaining elements of a set is called a *transposition*. π' can be obtained from π with the help of $k - 1$ transpositions $\vartheta_{i_j, i_{j+1}}$ as follows:

$$\pi' = \underbrace{\left(\pi \circ \vartheta_{i_1, i_2} \circ \vartheta_{i_2, i_3} \circ \vartheta_{i_3, i_4} \circ \ldots \circ \vartheta_{i_{k-2}, i_{k-1}}\right)}_{\alpha} \circ \vartheta_{i_{k-1}, i_k}$$

$$= \qquad\qquad\qquad \alpha \qquad\qquad\qquad \circ\ \vartheta_{i_{k-1}, i_k}$$

The transposition ϑ_{i_{k-1}, i_k} does not affect the sign of the product because its role in creating π' consists in interchanging $p(i_1)$ with the blank counter. Thus

$$\text{sign } \pi' = \text{sign } \alpha = (-1)^{k-2}\, \text{sign } \pi.$$

Since k is even, sign $\pi' = $ sign π. This implies that

(ii) paths of type (II) do not change the parity of the permutation corresponding to the previous arrangement of the counters.

According to (i) and (ii) the permutation π^* of the counters, which corresponds to the solution of the puzzle, has the same sign as the initial permutation π_0.

The permutation $\pi^* = \begin{pmatrix} 1 & 2 & \cdots & 16 \\ 1 & 2 & \cdots & \bigcirc \end{pmatrix}$ is even. Therefore a

necessary condition for the solution of the fifteen puzzle is that the initial arrangement of the counters corresponds to an even permutation.

Problem 127

Each step in the solution of Rubik's puzzle is a rotation ρ of a layer L of C through $\pm 90°$ (Fig. 2.77). ρ permutes cyclically the vertex cubelets and the edge cubelets of L, leaving the remaining cubelets fixed. Let Π' nd Π'' be arrangements of the cubelets before, and after the rotation ρ respectively. Denote by $\sigma_{\Pi'}(i_1)$, $\sigma_{\Pi'}(i_2)$, $\sigma_{\Pi'}(i_3)$, and $\sigma_{\Pi'}(i_4)$ the vertex cubelets of L before the rotation. The permutation $\sigma_{\Pi''}$, corresponding to Π'', is given by

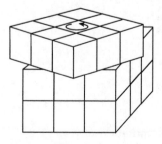

Fig. 2.77

$$\sigma_{\Pi''} = \begin{pmatrix} i_1 & i_2 & i_3 & i_4 & \dots & j\dots & \dots & k & \dots \\ \sigma_{\Pi'}(i_2) & \sigma_{\Pi'}(i_3) & \sigma_{\Pi'}(i_4) & \sigma_{\Pi'}(i_1) & \dots & \sigma_{\Pi'}(j) & \dots & \sigma_{\Pi'}(k) & \dots \end{pmatrix}$$

Thus

$$\sigma_{\Pi''} = \sigma_{\Pi'} \circ \vartheta_{1,2} \circ \vartheta_{2,3} \circ \vartheta_{3,4},$$

where $\vartheta_{i,j}$ is the transposition interchanging i and j (see the solution of Problem 126). This implies that

$$\text{sign } \sigma_{\Pi''} = (-1)^3 \text{ sign } \sigma_{\Pi'} = - \text{ sign } \sigma_{\Pi'}.$$

Similarly,

$$\text{sign } \tau_{\Pi''} = (-1)^3 \text{ sign } \tau_{\Pi'} = - \text{ sign } \tau_{\Pi'}.$$

Thus $\text{sign } \sigma_{\Pi''} \cdot \text{sign } \tau_{\Pi''} = \text{sign } \sigma_{\Pi'} \cdot \text{sign } \tau_{\Pi'}$; in other words the product of the signs of σ and τ remains invariant under each set of transformations during the game. In the final arrangement Π^*, representing the solution of the puzzle, σ_{Π^*} and τ_{Π^*} are identity permutations, so that $\text{sign } \sigma_{\Pi^*} \cdot \text{sign } \tau_{\Pi^*} = 1 \cdot 1 = 1$.

Therefore:

A necessary condition for the solvability of the puzzle is that $\text{sign } \sigma_{\Pi} \cdot \text{sign } \tau_{\Pi} = 1$ for the initial arrangement Π of the cubelets.

Problem 128

We shall study the changes of the angles $\alpha_{\Pi}(i)$ under rotations of the layers through $\pm 90°$ and $180°$. During this study the core of the cube C is held fixed, so that its blue and green pieces are horizontal (Fig. 2.78).

It is easy to check the truth of the following propositions.

(1) Rotations of horizontal layers through $\pm 90°$ and $180°$ leave $\alpha_{\Pi}(i)$ invariant for all $i = 1, \dots, 8$.

(2) Rotations of vertical layers through $180°$ leave $\alpha_{\Pi}(i)$ invariant for all $i = 1, \dots, 8$.

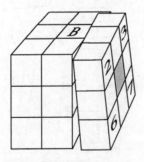

Fig. 2.78

(3) A rotation of an arbitrary vertical layer L through $\pm 90°$ transforms the angles $\alpha_\Pi(i_1)$ and $\alpha_\Pi(i_3)$ at two opposite corners of L into $\alpha_\Pi(i_1) + 120°$ and $\alpha_\Pi(i_3) + 120°$ respectively, the angles $\alpha_\Pi(i_2)$ and $\alpha_\Pi(i_4)$ at the two other corners of L into $\alpha_\Pi(i_2) + 240°$ and $\alpha_\Pi(i_4) + 240°$ respectively, and leaves the remaining angles $\alpha_\Pi(i)$ unaltered.

(1), (2), and (3) imply that the sum Σ'_Π of $\alpha_\Pi(1), \ldots, \alpha_\Pi(8)$, taken modulo 360°, remains invariant under each step in the process of solving the puzzle. In the final arrangement Π^*, when $\alpha_{\Pi^*}(i) = 0$ for all i, the sum $\Sigma'_{\Pi^*} = 0$. Thus:

The condition $\Sigma'_{\Pi^*} \equiv 0 \pmod{360°}$ is necessary for any initial arrangement Π of the cubelets which leads to a solution of Rubik's puzzle.

Problem 129

We shall study the changes of the angles $\beta_\Pi(j)$ under rotations of the layers through $\pm 90°$ and $180°$. The cube C is held in the same position as in Problem 128, with horizontal blue and green central pieces (Fig. 2.79).

It is easy to check on a model the truth of the following statement: C has a special pair of opposite layers L_1 and L_2 such that

(1) rotations of L_1 or L_2 through $\pm 90°$ transform the angles $B_\Pi(i)$ of the four cubelets $\tau_\Pi(i)$ of L_1 or L_2 respectively into $B_\Pi(i) + 180°$, and leave the remaining angles $\beta_\Pi(j)$ unaltered, and

(2) rotations of L_1 or L_2 through $180°$ and all rotations through $\pm 90°$ or $180°$ of the remaining layers leave the angles $\beta_\Pi(1), \beta_\Pi(2), \ldots, \beta_\Pi(12)$ unaltered.

From (1) and (2) it follows that $\Sigma''_\Pi \pmod{360°}$ is invariant under all transformations of C during the game. In the final arrangement Π^*, representing the solution of the game, $\beta_{\Pi^*}(j) = 0$ for all $j = 1, \ldots, 12$. This implies that $\Sigma''_{\Pi^*} = 0$. Hence,

$\Sigma''_{\Pi^*} \equiv 0 \pmod{360°}$ is a necessary condition for any initial arrangement Π of the cubelets leading to the solution of the puzzle.

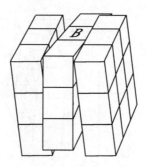

Fig. 2.79

Problem 130

(a) The vertex cubelets can be fitted into eight empty cells at the vertices of
     C in $8 \cdot 7 \cdot 6 \cdot \ldots \cdot 2 \cdot 1 = 8!$ different ways. Each vertex cubelet can be
     fitted into a given cell i in three different ways, corresponding to the
     three values $0°$, $120°$, and $240°$ of the angle $\alpha(i)$ (see Problem 128).
     Thus the number of ways in which the vertex cubelets can be arranged
     around the core of C is $V = 8! \cdot \underbrace{3 \cdot 3 \cdot \ldots \cdot 3}_{8 \text{ times}} = 8! \cdot 3^8$.

     The edge cubelets can be fitted into the 12 empty cells on the edges
     of C in $12 \cdot 11 \cdot 10 \cdot \ldots \cdot 2 \cdot 1 = 12!$ different ways. Each edge cubelet
     can be fitted into a given cell j in two different ways, corresponding to
     the two values $0°$ and $180°$ of the angle $\beta(j)$ (see Problem 129). Hence
     the number of ways in which the edge cubelets can be arranged
     around the core of C is $E = 12! \cdot \underbrace{2 \cdot 2 \cdot \ldots \cdot 2}_{12} = 12! \cdot 2^{12}$.

     Every fitting of the vertex cubelets into the vertex cells can be
     combined with every fitting of the edge cubelets into the edge cells.
     This implies that the total number of different ways in which C can be
     assembled from loose cubelets is

     $$V \cdot E = 8! \cdot 3^8 \cdot 12! \cdot 2^{12}.$$

(b) Let t be an arbitrary transformation of the assembled cube onto itself.
     t consists of a finite number of rotations of the layers of C through
     $k \cdot 90°$, where $k = 0, \pm 1, 2$. Denote by T the set of all transformations
     t of C. The set T forms a group under composition of trans-
     formations. Any $t \in T$ transforms an arbitrary arrangement Π of the
     cubelets into an arrangement $t(\Pi)$, and so preserves the three
     invariants: sign $\sigma_\Pi \cdot$ sign τ_Π, Σ'_Π and Σ''_Π. There are $2 \cdot 3 \cdot 2 = 12$
     possible combinations for the values of these invariants. For each
     combination we consider a corresponding arrangement Π_i of C from
     loose cubelets as shown in the table below. In particular, $\Pi_1 = \Pi^*$ is
     the basic arrangement with monochromatic faces.

Arrangement / Invariant	$\Pi_1 = \Pi^*$	Π_2	Π_3	Π_4	Π_5	Π_6	Π_7	Π_8	Π_9	Π_{10}	Π_{11}	Π_{12}
sign $\sigma \cdot$ sign τ	1	1	1	1	1	1	-1	-1	-1	-1	-1	-1
Σ'	$0°$	$120°$	$240°$	$0°$	$120°$	$240°$	$0°$	$120°$	$240°$	$0°$	$120°$	$240°$
Σ''	$0°$	$180°$	$0°$	$180°$	$0°$	$180°$	$0°$	$180°$	$0°$	$180°$	$0°$	$180°$

Consider the sets $S_i = \{t(\Pi_i); t \in \Pi\}$, $i = 1, \ldots, 12$. All elements of S_i have the same invariants as Π_i. Hence $S_i \cap S_j = \emptyset$ for $i, j = 1, 2, \ldots, 12$, $i \neq j$. For all $i = 1, 2, \ldots, 12$ the number $|S_i|$ of elements in S_i equals the number $|T|$ of elements in the group T. Thus

$$|S_1| + |S_2| + \ldots + |S_{12}| = 12|S_1| \leqslant V \cdot E,$$

i.e.

$$|S_1| \leqslant \tfrac{1}{12} \cdot 8! \cdot 3^8 \cdot 12! \cdot 2^{12} = 8! \cdot 3^8 \cdot 11! \cdot 2^{12}.$$

Since $\Pi_1 = \Pi^*$, the set S_1 consists of those arrangements, of the cubelets which lead to the solution of the puzzle. Therefore the set of the arrangements which does not lead to a solution is

$$V \cdot E - |S_1| \geqslant 8! \cdot 3^8 \cdot 12! \cdot 2^{12} - 8! \cdot 3^8 \cdot 11! \cdot 2^{12}$$

$$= 8! \cdot 3^8 \cdot 10! \cdot 11^2 \cdot 2^{12}.$$

Remark. It can be proved (see for example Dubrowski [35] that $V \cdot E - |S_1| = 8! \cdot 3^8 \cdot 10! \cdot 11^2 \cdot 2^{12}$.

Chapter X

Problem 131

We shall give two proofs of the formula

$$D_n = n! \left[1 - \frac{1}{1!} + \frac{1}{2!} + \ldots + (-1) \frac{1}{n!} \right].$$

(2.100)

Proof 1 is based on the method called the *principle of inclusion and exclusion*.

We shall exclude from the set σ_n of all permutations of $S = \{1, 2, \ldots, n\}$ those permutations which leave at least one element fixed. σ_n consists of $n!$ permutations. The number of permutations which leave at least one chosen element of S fixed is $(n-1)!$. There are $\binom{n}{1}$ ways of choosing an element of S. We subtract $\binom{n}{1}(n-1)!$ from the total of $n!$ permutations of σ_n.

The number $n! - \binom{n}{1}(n-1)!$ is smaller than D_n. Indeed, we have *excluded* each permutation with exactly k fixed elements $\binom{k}{1}$ times, for $k = 2, 3, \ldots$. We shall *include* them again by adding $\binom{n}{2}(n-2)!$. (There are $\binom{n}{2}$ choices of two elements from S, and for each choice of two elements there are $(n-2)!$ permutations fixing these elements.) The expression $n! - \binom{n}{1}(n-1)! + \binom{n}{2}(n-2)!$ counts each permutation fixing exactly k elements $1 - \binom{k}{1} + \binom{k}{2}$ times for $k = 3, 4, \ldots$. We shall *exclude* these permutations by subtracting $\binom{n}{3}(n-3)!$ from $n! - \binom{n}{1}(n-1)! + \binom{n}{2}(n-2)!$.

Proceeding in the above way we find that

$$D_n = n! - \binom{n}{1}(n-1)! + \binom{n}{2}(n-2)! - \ldots + (-1)^n \binom{n}{n}(n-n)!$$

$$= n! - \frac{n!}{1!(n-1)!}(n-1)! + \frac{n!}{2!(n-2)!}(n-2)! - \ldots$$

$$+ (-1)^n \frac{n!}{n!0!}(n-n)! = n! \left[1 - \frac{1}{1!} + \frac{1}{2!} - \ldots + (-1)^n \frac{1}{n!} \right].$$

273

Proof 2 uses a recurrence relation for D_n due to Euler. Let π be an arbitrary derangement of $S = \{1, 2, \ldots, n\}$, where $n \geqslant 3$. Suppose that $a_{2j} = j$ for a fixed $j \in \{2, \ldots, n\}$. Two cases can be distinguished:

(a) $a_{2j} = 1$. In this case the restriction of π to $\{2, 3, \ldots, j-1, j+1, \ldots, n\}$ is a derangement of $n - 2$ symbols. The number of these derangements is D_{n-2}.

(b) $a_{2j} \neq 1$. In this case the restriction of π to $S' = \{2, 3, \ldots, n\}$ is a one-to-one mapping φ of S' onto $S'' = \{1, \ldots, j-1, j+1, \ldots, n\}$ such that $\varphi(i) \neq i$ for $i \neq j$ and $\varphi(j) \neq 1$. The number of such mappings is equal to the number of derangements of $n - 1$ symbols, which is D_{n-1}.

(a) and (b) imply that there are $D_{n-2} + D_{n-1}$ derangements of S such that $a_{2j} = a_j$, j fixed. Since j can be chosen from $\{2, \ldots, n\}$ in $n - 1$ different ways, it follows that

$$D_n = (n-1)(D_{n-1} + D_{n-2}) \quad \text{for} \quad n \geqslant 3.$$

$D_1 = 0$ and $D_2 = 1$. Using these initial values and the above recurrence formula, one can prove the validity of (2.100) by induction.

Problem 132

Label the seats at the round table $1, 1', 2, 2', 3, 3', \ldots, n, n'$, and label the wives $1', 2', \ldots, n'$. Let i be the label of the husband of the wife labelled i' for $i = 1, \ldots, n'$.

Consider the situation when the i'th seat is occupied by the wife labelled i' for all $i = 1, 2, \ldots, n$. In that case for $i = 2, \ldots, n$ the husband labelled i cannot occupy the ith or the $(i-1)$th seat and the husband labelled 1 cannot occupy the first or the nth seat. Hence each admissible seating arrangement of the husbands corresponds to a normalized $3 \times n$ Latin square of the form

$$\begin{pmatrix} 1 & 2 & 3 & \ldots & n \\ n & 1 & 2 & \ldots & n-1 \\ a_{31} & a_{32} & a_{33} & \ldots & a_{3n} \end{pmatrix}$$

Conversely, each such Latin square corresponds to an admissible seating arrangement of the husbands. Thus the number of admissible seating arrangements of the husbands is the number of normalized $3 \times n$ Latin squares of type (1.52).

The same result holds for nay other fixed seating arrangement of the wives.

Problem 133

(a) Let $A(n,k)$ denote the number of ways of selecting k objects, no two consecutive, from n objects arranged in a row.

If $k = 1$, then $A_{n,1} = \binom{n}{1} = \binom{n-1+1}{1}$, and if $k = n > 1$ then

$$A_{n,n} = 0 = \binom{n-n+1}{n}$$

Consider now the case when $1 < k < n$. We distinguish between two types of selections:

(1) selections which do not include the first object in the row; their number is $A(n-1,k)$;

(2) selections which include the first object; by assumption, they cannot include the section object, therefore their number is $A(n-2, k-1)$.

Thus we have the recurrence formula

$$A(n,k) = A(n-1,k) + A(n-2, k-1).$$

Using this recurrence relation we can prove by induction that $A(n,k) = \binom{n-k+1}{k}$. Indeed, according to the induction hypothesis,

$$A(n-1,k) = \binom{n-k}{k} \quad \text{and} \quad A(n-2, k-1) = \binom{n-k}{k-1}.$$

Hence

$$A(n,k) = \binom{n-k}{k} + \binom{n-k}{k-1} = \binom{n-k+1}{k}.$$

(b) Let $B(n,k)$ be the number of ways of selecting k objects, no two consecutive, from n objects arranged in a circle.

We shall label the objects in the circle $1, 2, 3, \ldots, n$ (Fig. 2.80). Again we shall distinguish between two types of selections:

(1) selections which do not include the first object; their number is $A(n-1,k)$;

(2) selections which include the first object; they cannot include objects 2 and n, therefore their number is $A(n-3, k-1)$.

Hence

$$B(n,k) = A(n-1,k) + A(n-3, k-1)$$

$$= \binom{n-1-k+1}{k} + \binom{n-3-(k-1)+1}{k-1}$$

$$= \frac{n}{n-k}\binom{n-k}{k} \quad \text{for} \quad n > k.$$

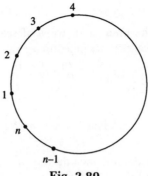

Fig. 2.80

We are now ready to reproduce Kaplansky's proof of the formula for the ménage numbers U_n (defined on p. 106).

U_n is the number of permutations π of $\{1, 2, \ldots, n\}$ which satisfy the conditions $\pi(i) \neq i, i-1$ for $i = 2, 3, \ldots, n$ and $\pi(1) \neq 1, n$. U_n will be determined by the principle of exclusion and inclusion (described in the solution of Problem 131).

We shall say that a permutation of $\{1, 2, \ldots, n\}$ has property p_i or p_i' if it maps i onto i, or onto $i-1$ respectively for $i = 2, \ldots, n$. A permutation has property p_1 or p_1' if it maps 1 onto 1, or onto n respectively. Our next aim is to determine the number of permutations of $\{1, \ldots, n\}$ which satisfy at least k of the properties p_i and p_j'. In Fig. 2.81 the labels of the properties $p_1', p_1, p_2', p_2, \ldots, p_n', p_n$ are arranged one after another in a circle. Clearly, a

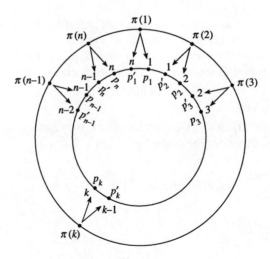

Fig. 2.81

permutation cannot have at the same time two consecutive properties of this arrangement. Hence

The number of choices of k properties from the set $p'_1, p_1, \ldots, p'_n, p_n$ is $B(2n, k)$ (see the solution of Problem 133(b)).

Let $\Pi(i_1, i_2, \ldots, i_k)$ be the set of those permutations of $\{1, 2, \ldots, n\}$ which satisfy k properties with suffixes $i_1, i_2, \ldots, i_k$. All permutations of $\Pi(i_1, i_2, \ldots, i_k)$ map $i_1, i_2, \ldots, i_k$ onto the same images $\Pi(i_1)$, $\Pi(i_2), \ldots, \Pi(i_k)$. The elements of $\{1, \ldots, n\}$, different from $i_1, \ldots, i_k$ are mapped by all permutations of $\Pi(i_1, i_2, \ldots, i_k)$ onto those elements of $\{1, \ldots, n\}$ which are different from $\Pi(i_1), \ldots, \Pi(i_k)$; this can be done in $(n - k)!$ different ways. Thus

There are $(n - k)! B(2n, k) = 2n/(2n - k) \binom{2n - k}{k} (n - k)!$ permutations of $\{1, \ldots, n\}$ that satisfy at least k of the properties p_i and p'_j.

Applying the principle of exclusion and inclusion, we find that

$$U_n = n! - \frac{2n}{2n - 1} \binom{2n - 1}{1} (n - 1)! + \frac{2n}{2n - 2} \binom{2n - 2}{2} (n - 2)! - \ldots$$
$$+ (-1)^n \cdot \frac{2n}{n} \binom{n}{n} 0!$$

Remark. The number of ways of seating n married couples at a round table with men and women in alternate positions so that no wife sits next to her husband is $2n! U_n$.

Proof. In view of Problem 132, if the i'th wife occupies the i'th seat for $i = 1, 2, \ldots, n$, then there are U_n admissible seating arrangements for the husbands. However, there are $n!$ ways in which the n wives can occupy the seats $1', 2', \ldots, n'$; moreover, there are $n!$ ways in which the wives can occupy the seats $1, 2, \ldots, n$. Thus the total number of seating arrangements such that no husband sits next to his wife is $2n! U_n$.

Problem 134

Denote the identity element of G by g_1, and the inverse of any element g_i of G by g_i^{-1}. The operation table of G is shown below.

$$
\begin{array}{ccccc}
g_1 \circ g_1 & g_1 \circ g_2 & \cdots & g_1 \circ g_j & \cdots & g_1 \circ g_n \\
g_2 \circ g_1 & g_2 \circ g_2 & \cdots & g_2 \circ g_j & \cdots & g_2 \circ g_n \\
g_3 \circ g_1 & g_3 \circ g_2 & \cdots & g_3 \circ g_j & \cdots & g_3 \circ g_n \\
\cdot & \cdot & \cdots & \cdot & \cdots & \cdot \\
g_i \circ g_1 & g_i \circ g_2 & \cdots & g_i \circ g_j & \cdots & g_i \circ g_n \\
\cdot & \cdot & \cdots & \cdot & \cdots & \cdot \\
g_n \circ g_1 & g_n \circ g_2 & \cdots & g_n \circ g_j & \cdots & g_n \circ g_n
\end{array}
$$

Our aim is to show that each row and each column of the table consists of the same set of n different elements.

Suppose that the ith row has two equal entries, say

$$g_i \circ g_j = g_i \circ g_k \quad \text{for} \quad j \neq k.$$

This implies that

$$g_i^{-1} \circ (g_i \circ g_j) = g_i^{-1} \circ (g_i \circ g_k)$$

$$\underbrace{(g_i^{-1} \circ g_i)}_{g_1} \circ g_j = \underbrace{(g_i^{-1} \circ g_i)}_{g_1} \circ g_k$$

Hence

$$g_j = g_k,$$

which is a contradiction. Thus the ith row consists of n different elements of G. Since the row has n elements, it contains each element of G exactly once.

Similarly, one can show that for each $j = 1, \ldots, n$ the jth column consists of n elements of G.

Hence the operation table of a finite group G of n elements is an $n \times n$ Latin square.

Problem 135

Figure 2.82(a) shows the first few entries in the construction of an infinite square grid S (bordered from the top, and from the left) such that each row and each column of S contains every natural number exactly once. The stepwise construction of S is illustrated in Fig. 2.82(b). We start from a 1×1 square $A_1 = \boxed{1}$. At the nth stage the square A_n is assembled from two copies of A_{n-1} on the main diagonal of A_n, and from two copies of the square labelled $A_{n-1} + 2^{n-2}$. The square $A_{n-1} + 2^{n-2}$ is obtained from A_{n-1} by adding 2^{n-2} to each of its entries.

The above construction process is performed for $n = 1, 2, 3, \ldots$. In this way an infinite square is obtained in which the first row and the first column represent the sequence $1, 2,, \ldots$ of all natural numbers; the

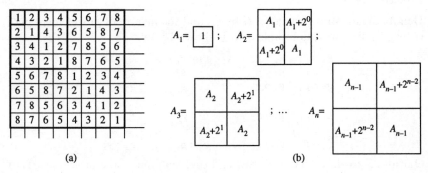

(a)   (b)

Fig. 2.82

remaining rows and columns of S are permutations of this sequence, different from one another.

Problem 136

Without loss of generality we shall assume that $A_1, \ldots, A_k$ are *normalized*, mutually orthogonal Latin squares:

$$A_i = \begin{pmatrix} 1 & 2 & \ldots\ldots & n \\ a_{21}^{(i)} & a_{22}^{(i)} & & a_{2n}^{(i)} \\ \cdot & \cdot & \ldots\ldots & \cdot \\ a_{n1}^{(i)} & a_{n2}^{(i)} & \ldots\ldots & a_{nn}^{(i)} \end{pmatrix} \quad \text{for} \quad i = 1, \ldots, k.$$

$a_{21}^{(i)} \neq 1$ for $i = 1, \ldots, k$, since no column of a Latin square can contain two equal entries. The numbers $a_{21}^{(1)}, a_{21}^{(2)}, \ldots, a_{21}^{(k)}$ must be different. Indeed, $a_{21}^{(i')} = a_{21}^{(i'')}$ for some $1 \leqslant i' < i'' \leqslant k$ would imply that the Latin squares A_i' and A_i'' are not orthogonal. Since $a_{21}^{(i)} \in \{2, 3, \ldots, n\}$ for $i = 1, \ldots, k$, this implies that $k \leqslant n - 1$.

Problem 137

The definition of $\mathcal{A}_5$ can be extended to $\mathcal{A}_p$ for any prime p.

$\mathcal{A}_p$ consists of points and lines. The points of $\mathcal{A}_p$ are the ordered pairs (x, y) of elements of $\mathbb{Z}_p = \{0, 1, 2, \ldots, p - 1\}$. A line of $\mathcal{A}_p$ is a set of points whose 'coordinates' x, y satisfy an equation of type (I)

$$y = m \otimes x \oplus n, \quad m, n \in \mathbb{Z}_p,$$

or of type (II)

$$x = a \quad , \quad a \in \mathbb{Z}_p.$$

$\oplus$ and $\otimes$ denote addition, respectively multiplication modulo p. $\mathbb{Z}_p$ is a field with respect to $\oplus$ and $\otimes$, that is:

$\mathbb{Z}_p$ is a group with respect to $\oplus$, and $\mathbb{Z}_p^* = \{1, 2, \ldots, p - 1\}$ is a group with respect to $\otimes$ (see Problem 112). Moreover, $(m \oplus n) \otimes k = (m \otimes k) \oplus (n \otimes k)$, and $k \otimes (m \oplus n) = (k \otimes m) \oplus (k \otimes n)$ for all $m, n, k \in \mathbb{Z}_p$.

The proof that $\mathcal{A}_p$ is an affine plane relies on the properties of the field $\mathbb{Z}_p$. We have to verify axioms $\mathbb{A}_1$, $\mathbb{A}_2$, and $\mathbb{A}_3$ for A_p. This will be done in three steps.

(i) $\mathcal{A}_p$ satisfies axiom $\mathbb{A}_1$.

Proof. Let $P_1 = (x_1, y_1)$ and $P_2 = (x_2, y_2)$ be two distinct points of $\mathcal{A}_p$. Two cases will be distinguished:

Case 1. $x_1 \neq x_2$. In this case P_1 and P_2 cannot belong to a common line of type (II). P_1 and P_2 belong to a common line of type (I) if and only if there are numbers $m, n \in \mathbb{Z}_p$ such that

and
$$\left. \begin{array}{l} y_1 = m \otimes x_1 \oplus n \\ \\ y_2 = m \otimes x_2 \oplus n. \end{array} \right\} \tag{2.101}$$

Since $\mathbb{Z}_p (\oplus, \otimes)$ is a field, the above system of simultaneous equations yields

$$y_1 \oplus (-y_2) = m \otimes (x_1 \oplus (-x_2)), \tag{2.102}$$

where $-y_2$ and $-x_2$ are the inverses of y_2 and x_2 respectively, with respect to the operation $\oplus$.

From $x_1 \neq x_2$ it follows that $x_1 \oplus (-x_2) \neq 0$. Hence there is a unique element m satisfying the equations (2.101). Also there is a unique n satisfying (2.101). Thus P_1 and P_2 are on a unique line of type (I).

Case 2. $x_1 = x_2$. Then P_1 and P_2 are on a unique line of type (II). This line has the equation $x = a$, where $a = x_1 = x_2$.

Suppose that P_1 and P_2 are also on a line of type (I). In that case their coordinates satisfy the system of simultaneous equations (2.101). Hence there exists a solution m of (2.102). However, the right-hand side of this equation is $m \otimes 0 = 0$, while the left-hand side of the equation is $y_1 \oplus (-y_2) \neq 0$ (since $P_1 \neq P_2$). This contradiction shows that there is no line of type (I) containing P_1 and P_2.

Cases (1) and (2) imply that

> any two distinct points $\mathcal{A}_p$ are on a unique line of $\mathcal{A}_p$.

This proves (i).

(ii) $\mathcal{A}_p$ satisfies axiom $\mathbb{A}_2$.

Proof. First we shall study the conditions under which two distinct lines of $\mathcal{A}_p$ have a point in common.

Let ℓ_1 and ℓ_2 be two distinct lines of $\mathcal{A}_p$. Three cases will be distinguished.

Case 1. ℓ_1 and ℓ_2 are of type (II). In this case their equations are $x = a_1$ and $x = a_2$ respectively, with $a_1, a_2 \in \mathbb{Z}_p$, $a_1 \neq a_2$. Thus ℓ_1 and ℓ_2 have no point in common.

Case 2. ℓ_1 is a line of type (I) with equation $y = m \otimes x \oplus n$, and ℓ_2 is a line of type (II) with equation $x = a$ for some $m, n, a \in \mathbb{Z}_p$. In this case $(a, m \otimes a \oplus n)$ is the only point common to ℓ_1 and ℓ_2.

Case 3. ℓ_i are of type (I) with equations $y = m_i \otimes x \oplus n_i$, where $m_i, n_i \in \mathbb{Z}_p$, $i = 1, 2$. If ℓ_1 and ℓ_2 have a common point P, then the coordinates x, y of P are solutions of the system of simultaneous equations

$$\left. \begin{array}{l} y = m_1 \otimes x \oplus n_1 \\ y = m_2 \otimes x \oplus n_2. \end{array} \right\} \tag{2.103}$$

(2.103) yields

$$(m_1 \oplus (-m_2)) \otimes x = n_2 \oplus (-n_1). \tag{2.104}$$

Equation (2.104) has a unique solution if and only if $m_1 \oplus (-m_2) \neq 0$. In that case there is a unique value of y satisfying both equations in (2.103). Thus two distinct lines of type (I) have a unique point in common if and only if $m_1 \neq m_2$. Otherwise the lines do not meet.

To sum up:

Two distinct lines of $\mathcal{A}_p$ have a point in common if and only if:

they are of different types, or

they are both of type (I) with equations $y = m_i \otimes x + n_i$, $i = 1, 2$, with $m_1 \neq m_2$.

We are now able to prove the validity of axiom $\mathbf{A}_2$:

Let $P = (x_0, y_0)$ be a point of $\mathcal{A}_p$, and let ℓ be a line of $\mathcal{A}_p$, not containing P.

If ℓ is a line of type (I) with equation $y = m \otimes x \oplus n$, then there is a unique line ℓ' through P not meeting ℓ. The equation of ℓ' is

$$y = m \otimes x \oplus n_0, \quad \text{where} \quad n_0 = -(m \otimes x_0) \oplus y_0.$$

If ℓ is a line of type (II) with equation $x = a$, then there is a unique line ℓ' through P not meeting ℓ. Its equation is

$$x = x_0.$$

Hence $\mathcal{A}_p$ satisfies axiom $\mathbf{A}_2$.

(iii) $\mathcal{A}_p$ satisfies axiom $\mathbf{A}_3$ because $\mathcal{A}_p$ contains three non-collinear points (for example $(0, 0)$, $(0, 1)$, and $(1, 0)$).

Thus we have completed the proof of parts (a) and (b) of the problem:

$\mathcal{A}_p$ is an affine plane for any prime number p.

(c) The structure $\mathcal{A}_6$ consists of points (x, y), $x, y \in \mathbb{Z}_6$, and of lines with equations $y = m \otimes x \oplus n$, or $x = a$, for $m, n, a \in \mathbb{Z}_6$. The symbols $\oplus$ and $\otimes$ denote addition, respectively multiplication modulo 6.

It is easy to show that $\mathcal{A}_6$ does not satisfy all axioms of an affine plane. For example, there is no line through the points $P_1(1, 2)$ and $P_2(3, 5)$; this is because

$$1 \neq 3 \text{ (hence } P_1 \text{ and } P_2 \text{ are not on a line of type II),}$$

and the system

$$5 = m \otimes 3 \oplus n,$$
$$2 = m \otimes 1 \oplus n,$$

analogous to (2.101), yields the equation

$$3 = m \otimes 2$$

which has no solution in $\mathbb{Z}_6$.

Thus axiom $\mathbf{A}_1$ does not hold in $\mathcal{A}_6$.

Problem 138

(a) Let ℓ be a line of $\mathcal{A}$ with n points $A_1, A_2, \ldots, A_n$. According to axiom $\mathbf{A}_3$ there is a point $P \in \mathcal{A}, P \notin \ell$. The point A_i and P belong to a unique line ℓ_i, $i = 1, \ldots, n$ (by axiom $\mathbf{A}_1$). In view of axiom $\mathbf{A}_2$, there is a unique line $\ell_{n+1} \ni P$ such that ℓ and ℓ_{n+1} have no point in common. This implies that $\ell_1, \ell_2, \ldots, \ell_n, \ell_{n+1}$ are the only lines containing P (Fig. 2.83).

We shall now prove that an arbitrary line ℓ' of $\mathcal{A}$ consists of n points. Three cases will be distinguished.

Case 1. $\ell' \not\ni P$. In that case, by axiom $\mathbf{A}_2$, exactly one of the lines $\ell_1, \ell_2, \ldots, \ell_{n+1}$ has no point in common with ℓ'. The remaining lines $\ell_i \ni P$ intersect ℓ' in n distinct points. These are all the points of ℓ'. This is so because an $(n+1)$st point on ℓ' would determine an $(n+2)$nd line containing P, which is not possible.

Case 2. $\ell' \ni P$, $\ell' \neq \ell_{n+1}$. In that case, $\ell' = \ell_i$ for some $i = 1, 2, ., n$. Let $\ell_j \ni P$ be a line different from ℓ_i. By axiom $\mathbf{A}_2$, there is a unique line $\ell'_k \ni A_k$ such that $\ell'_k \cap \ell_j = \emptyset$ for each $k = 1, \ldots, n, k \neq j$. The line ℓ'_k meets ℓ_i at a point C_k for $k = 1, \ldots, n, k \neq i, j$. Otherwise P would belong to two lines, ℓ_j and ℓ_i,

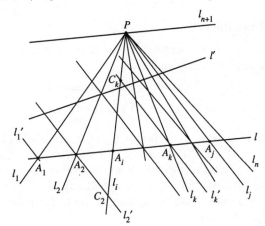

Fig. 2.83

not intersecting ℓ'_k. But this would contradict axiom $\mathbb{A}_2$. Hence ℓ_i contains at least n points: the $n-2$ points C_k and the points P and A_i. These are all the points of ℓ_i. (The existence of a further point $D \in \ell_i$ would imply the existence of a line $\bar{\ell} \ni D$ such that $\bar{\ell} \cap \ell_j = \emptyset$. The intersection $\bar{\ell} \cap \ell$ would represent an $(n+1)$st point of ℓ, contradicting our assumption that ℓ of n points.)

Case 3. $\ell' = \ell_n + 1$. By $\mathbb{A}_2$, for each $k = 2, \ldots, n$ there is a unique line $\ell'_k \ni A_k$ such that $\ell'_k \cap \ell_1 = \emptyset$. Each ℓ'_k meets ℓ_{n+1} at a point, B_k say, for otherwise A_k belongs to two lines, ℓ'_k and ℓ, which do not meet ℓ_{n+1}, which contradicts $\mathbb{A}_2$. Hence ℓ' ($\ell' = \ell_{n+1}$) contains at least the n points P and B_k ($k = 2, \ldots, n$). These are all the points of ℓ'. (The existence of a further point $D \in \ell'$ would imply that there is a line $\bar{\ell}$ through D not meeting ℓ_1; this line would meet ℓ_1 giving an $(n+1)$st point on ℓ, contradicting our assumption that ℓ consists of n points.) Thus ℓ' contains n points.

Thus every line of $\mathcal{A}$ contains n points.

(b) Let Q be an arbitrary point of $\mathcal{A}$. There is a line q in $\mathcal{A}$ such that $q \not\ni Q$. Axiom $\mathbb{A}_1$ implies that the n points $R_1, \ldots, R_n$ of q determine n lines 'joining' them to Q. According to $\mathbb{A}_2$, there is a unique line containing Q which has no point in common with q.

Thus there are exactly $n+1$ lines containing Q.

(c) Let P be an arbitrary point of $\mathcal{A}$. Any point of $\mathcal{A}$ different from P belongs to exactly one of the $n+1$ lines through P. Each of these lines contains $n-1$ points different from P.

Thus the number of points of $\mathcal{A}$ is $1 + (n+1)(n-1) = n^2$.

(d) Any two distinct points of $\mathcal{A}$ are on a unique line of $\mathcal{A}$. The number of distinct pairs of points in $\mathcal{A}$ is $\binom{n^2}{2} = n^2(n^2-1)/2$. However, this number does not correspond to the number of lines in $\mathcal{A}$ because each line ℓ is determined by $\binom{n}{2} = n(n-1)/2$ pairs of points of ℓ. Thus the number of lines in $\mathcal{A}$ is $(n^2(n^2-1)/2)/(n(n-1)/2) = n(n+1)$.

(e) We start with the following observation:

(∗) If a line ℓ has no common point with two distinct lines ℓ' and ℓ'', then $\ell' \cap \ell'' = \emptyset$.

Indeed, $\ell' \cap \ell'' = R$ would imply that R belongs to two lines ℓ' and ℓ'' which have no common point with ℓ, contradicting axiom $\mathbb{A}_2$.

Now let us consider an arbitrary line ℓ of $\mathcal{A}$. Let ℓ' be a line meeting ℓ at some point P (Fig. 2.84). Denote the remaining points of ℓ' by $P_1, P_2, \ldots, P_{n-1}$. According to axiom $\mathbb{A}_2$, through each $P_i, i = 1, 2, \ldots, n-1$, there passes a unique ℓ_i such that $\ell \cap \ell_i = \emptyset$. In

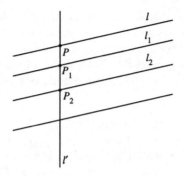

Fig. 2.84

view of (∗) it follows that $\ell_i \cap \ell_j = \emptyset$ for all $i,j = 1,\ldots,n-1$. Thus the points of $\ell, \ell_1, \ell_2, \ldots, \ell_{n-1}$ are all different; their number is $n \cdot n - n^2$, which is the number of all points in $\mathcal{A}$. This proves that

ℓ belongs to a unique set of n lines no two of which have a point in common.

Problem 139

(a) For any $i \in \{1, \ldots, n\}$ the ith line of $\overline{R}$ is parallel only to the lines of the parallel class $\overline{R}$ (see Problem 138(e)). Hence line i intersects an arbitrary line $j, j = 1, \ldots, n$, of the parallel class $\overline{C}$. The intersection of the lines i and j is the point (i, j) of $\mathcal{A}$.

(b) In view of Problem 138(e), each point of $\mathcal{A}$ belong to a unique line of each parallel class. Hence there is exactly one line of each parallel class $\overline{k}$ through (i, j). (Fig. 2.85).

(c) (i)   For each $i,j \in \{1,..,n\}$ denote the entry in the ith row and jth column of A_k by $a_{ij}^{(k)}$. According to the construction of A_k, the entry $a_{ij}^{(k)}$ is the label of the line of the parallel class $\overline{k}$ which passes through the point (i, j).

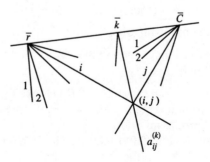

Fig. 2.85

Two distinct point (i, j) and (i, j') belong to the ith line of $\overline{R}$. Thus $t = a_{ij}^{(k)} = a_{ij'}^{(k)}$, for $j \neq j'$ would imply that the tth line of $\overline{k}$ meets the line i of $\overline{R}$ in *two* different points. This contradiction shows the each row of A_k consists of n different entries; these are the elements of the set $\{1, 2, \ldots, n\}$.

Similarly, $a_{ij}^{(k)} = a_{i'j}^{(k)}$ for $i \neq i'$ would imply that the jth line of $\overline{C}$ is met by the same line of $\overline{k}$ in two different points, contradicting axiom $\mathbf{A}_1$. Hence each column of A_k consists of the n different entries $1, 2, \ldots, n$.

Thus A_k is a Latin square for $k = 1, 2, \ldots, n$.

(ii) Let A'_k and $A_{k''}$ be two latin squares, $k', k'' \in \{1, \ldots, n-1\}$, $k' \neq k''$. If $a_{ij}^{(k')} = t'$ and $a_{ij}^{(k'')} = t''$ for the same $i, j \in \{1, \ldots, n\}$, this means that the line t' of the parallel class $\overline{k}'$ and the line t'' of the parallel class $\overline{k}''$ meet at the point (i, j) of A. If $a_{ij}^{(k')} = a_{pq}^{(k')} = t'$ and $a_{ij}^{(k'')} = a_{pq}^{(k'')} = t''$ for $i, j, p, q \in \{1, \ldots, n\}$, $(i, j) \neq (p, q)$, then the lines t' and t'' would meet at two distinct points (i, j) and (p, q), which contradicts axiom $\mathbf{A}_1$. This implies that the ordered pairs of entries $\left(a_{ij}^{(k')}, a_{ij}^{(k'')} \right)$, $i,, \in \{1, \ldots, n\}$ are all different.

Thus A'_k and A''_k are orthogonal Latin squares.

The mutually orthogonal Latin squares $A_1, A_2, \ldots, A_{n-1}$ form a complete set because their number is $n - 1$.

Problem 140

Let $A_1 = \left(a_{ij}^{(1)} \right)$, $A_2 = \left(a_{ij}^{(2)} \right), \ldots, A_{n-1} = \left(a_{ij}^{(n-1)} \right)$ be a complete set of orthogonal Latin squares with entries $1, 2, ., n$. We shall construct a structure $\mathcal{A}$ of points and lines, based on these Latin squares, as follows.

The points of $\mathcal{A}$ are the ordered pairs (i, j) of numbers $i, j \in \{1, 2, \ldots, n\}$. Thus $\mathcal{A}$ has n^2 points.

The lines are the following point sets:

(I) $\ell_i^{(\overline{R})} = \{(i, 1), (i, 2), \ldots, (i, n)\}$, $i = 1, \ldots, n$;

(II) $\ell_j^{(\overline{C})} = \{(1, j), (2, j), \ldots, (n, j)\}$, $j = 1, \ldots, n$;

(III) $\ell_t^{(k)} = \{(i, j); a_{ij}^{(k)} = t\}$, $t = 1, \ldots, n$; $k = 1, \ldots, n-1$.

$\mathcal{A}$ has n lines of type (I), n lines of type (II), and $n(n - 1)$ lines of type (III). Altogether there are $n^2 + n$ lines in $\mathcal{A}$.

Since the arrays A_k are Latin squares, it follows that all lines of $\mathcal{A}$ consist of n points, and that each line of type (III) has exactly one point in common with any line of type (I) and of type (II).

Any line of type (I) meets any line of type (II) in exactly one point. Lines of type (I) form a 'parallel class' $\overline{R}$, because no two lines of $\overline{R}$ have a point in common. Similarly, the lines of type (II) form a parallel class $\overline{C}$, and the lines of type (III) form $n-1$ parallel classes $\bar{k} = \{\ell_1^{(k)}, \ell_2^{(k)}, \ldots, \ell_n^{(k)}\}$ for $k = 1, 2, \ldots, n-1$.

Any two lines from different classes $\bar{k}'$ and $\bar{k}''$ have a unique point in common. This follows from the fact that the corresponding Latin squares $A_{k'}$ and $A_{k''}$ are orthogonal.

Each point of $\mathcal{A}$ belongs to $n+1$ lines.

From the above statements it is easy to deduce that $\mathcal{A}$ satisfies axioms $\mathbb{A}_1$, $\mathbb{A}_2$, and $\mathbb{A}_3$ of an affine plane. The reader is asked to verify this claim.

Problem 141

Each parallel class of $\mathcal{A}$ is a point of $\mathcal{A}^*$; let us call such a point an *ideal point* of $\mathcal{A}^*$. The set of all ideal points will be called the *ideal line* of $\mathcal{A}^*$. The points and lines of $\mathcal{A}^*$ have the following properties.

(a) If P and Q are two distinct points of $\mathcal{A}^*$, then the following cases can be distinguished.

  (i) P and Q are points of $\mathcal{A}$; in that case there is a unique line of $\mathcal{A}$ containing P and Q; P and Q do not belong to the ideal line.

  (ii) P is a point of $\mathcal{A}$ and Q is an ideal point; in that case P lies on a unique line of $\mathcal{A}$ which belongs to the parallel class Q; P and Q are not connected by the ideal line.

  (iii) P and Q are both ideal points, in which case they belong to the ideal line, but to no line of $\mathcal{A}$.

  We conclude that $\mathcal{A}^*$ satisfies $\mathbb{P}_1$.

(b) Let ℓ_1 and ℓ_2 be two distinct lines of $\mathcal{A}^*$.

  (i)   If ℓ_1 and ℓ_2 belong to $\mathcal{A}$, and are not parallel in $\mathcal{A}$, then they meet at a unique point of $\mathcal{A}$. ℓ_1 and ℓ_2 have no common ideal point.

  (ii) If ℓ_1 and ℓ_2 belong to $\mathcal{A}$, and are parallel in $\mathcal{A}$, then they belong to the same parallel class of $\mathcal{A}$. The ideal point of $\mathcal{A}^*$ representing this parallel class is a common point of ℓ_1 and ℓ_2.

  (iii) If ℓ_1 is the ideal line, then ℓ_2 belongs to $\mathcal{A}$ and ℓ_1 and ℓ_2 meet at the ideal point which is the parallel class containing ℓ_2.

  We conclude that $\mathcal{A}^*$ satisfies $\mathbb{P}_2$.

(c) $\mathcal{A}$ contains three non-collinear points P_1, P_2, and P_3. Through P_1 there is a line ℓ_1 parallel to the line containing P_2 and P_3. Similarly, through P_2 there is a line ℓ_2 parallel to the line containing P_1 and P_3. The lines ℓ_1 and ℓ_2 belong to distinct parallel classes of $\mathcal{A}$. Hence ℓ_1 and ℓ_2 meet at a common point P_4 of $\mathcal{A}$. The points P_1, P_2, P_3, and P_4 are four points of $\mathcal{A}^*$ such that no three of them belong to a line of $\mathcal{A}^*$.

This proves that $\mathcal{A}^*$ satisfies $\mathbb{P}_3$.

In view of (a), (b), and (c), $\mathcal{A}^*$ is a projective plane.

Problem 142

(a) Any two distinct points P, Q of π_ℓ are on a unique line of π different from ℓ. Thus

P, Q are on a unique line of π_ℓ.

(b) Let ℓ' be a line of π_ℓ and P a point of π_ℓ not on ℓ'. The lines ℓ' and ℓ meet at a unique point P' in π, and P and P' are joined by a unique line ℓ'' of π. The lines of π_ℓ through P, different from ℓ'', meet ℓ' at points not on ℓ. Hence,

in π_ℓ there exists a unique line ℓ'' through P which has no point in common with ℓ'.

(c) π contains four points P_1, P_2, P_3, and P_4 no three of which are collinear. If three of these points, say P_1, P_2, P_3, do not belong to ℓ, then P_1, P_2, P_3 are three points of π_ℓ, not on a line of π_ℓ. Suppose that P_3 and P_4 belong to ℓ. The line through P_1 and P_4 meets the line through P_2 and P_3 at a unique point P_5 in π. The point P_5 is not on ℓ. Hence

P_1, P_2, and P_5 are three points of π_ℓ not on a common line of π_ℓ.

From (a), (b), and (c) it follows that π_ℓ is an affine plane.

Problem 143

Proof by induction.

If $n = 1$ then the plane is divided into two half-planes by a single line ℓ_1. The half-planes are the countries of a map M_1 that can be properly coloured, using two colours.

Suppose that any map M_n, formed by n lines in a plane, can be properly coloured using two colours.

Now consider a map M_{n+1} formed by $n+1$ lines in a plane. Let us remove a line ℓ from M_{n+1}. The remaining lines form a map M_n^* with countries $R_1, R_2, \ldots, R_k$. According to the induction hypothesis, M_n^* can be properly coloured with two colours, say black and white. Let C^* be such a colouring of M_n^* (Fig. 2.86(a)). Then we return ℓ to its original position. ℓ divides the plane into two half-planes π_1 and π_2. The countries of the map M_{n+1} are $R_i \cap \pi_1$ and $R_i \cap \pi_2$, $i = 1, 2, \ldots, k$. We shall now construct a colouring C of M_{n+1} as follows: the colours of the countries $R_i \cap \pi_1$ are left unchanged, while the black or white colours of $R_i \cap \pi_2$ are changed to white or black respectively (Fig. 2.86(b)). C is a proper colouring of M_{n+1}.

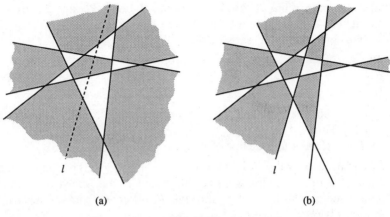

(a)   (b)

Fig. 2.86

This finishes the proof of the statement.

Problem 144

The statement can be proved by induction on n (cf. Problem 143). Another ingenious proof is described in Dynkin and Uspenskii [37], as follows.

Each country of the map is labelled with the number of circles surrounding it (Fig. 2.87). Our aim is to prove that

⊛ any two countries R_i, R_j with common boundary have labels differing by 1.

Indeed, the common boundary is the arc of a circle c which separates R_i and R_j: one of them is surrounded by c, the other not. Any other circle forming the map contains both R_i and R_j or neither. This proves ⊛.

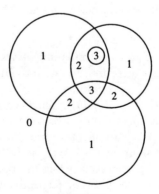

Fig. 2.87

We shall colour the countries with odd labels white and those with even labels black. According to $\circledast$, countries with a common boundary have different colours. Thus the colouring is proper.

Problem 145

(a) Suppose that a map M is properly coloured with two colours. Let P be an arbitrary vertex of M. Denote by $R_1, R_2, \ldots, R_k$ the countries with vertex P such that R_i and R_{i+1} are separated by a boundary b_i, $i = 1, 2, \ldots, k - 1$, and R_k and R_1 are separated by b_k (Fig. 2.88(a)). The colours of $R_1, R_2, \ldots, R_k$ alternate, and the colours of R_k and R_1 differ.

Hence the number k of the boundaries meeting at P is even.

(b) Let the number of boundaries meeting at each vertex of a map be even. We shall show that the maps can be properly coloured with two colours, using induction on the number n of all boundaries of the map.

The statement is true for a map with two boundaries (Fig. 288(b)).

Suppose that the statement is true for any map in which an even number of boundaries meet at each vertex, and in which the total number of boundaries does not exceed n.

Let M be a map with $n + 1$ boundaries, such that each of its vertices belongs to an even number of boundaries. This condition implies that, starting from an arbitrary vertex, say P_1 of M, we can traverse a sequence s of boundaries and eventually return to P_1 (the number of vertices in M is finite!). s is a closed path which does not intersect itself (Fig. 2.88(c)). By removing s from M a new map M^* is created. M^* has fewer than $n + 1$ boundaries. Moreover, the number of boundaries through any vertex P in M^* is the same as it was in M (if in M the vertex P did not belong to s), or two less than it was in M (if

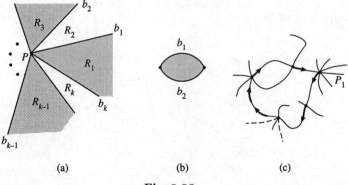

(a)   (b)   (c)

Fig. 2.88

in M the vertex P did belong to s). According to the induction hypothesis, M^* can be properly coloured with two colours. After doing this, we restore s and interchange the two colours inside (or outside) s. This yields a proper colouring of M with two colours.

Problem 146

We shall prove Euler's theorem by induction on the number of boundaries.

If a map has one vertex, hence once country and zero boundaries, then $1 - 0 + 1 = 2$.

Suppose that the formula $v - e + r = 2$ is satisfied for any map with $e = n$ boundaries.

Consider a map M with $n + 1$ boundaries, v vertices and r countries. We distinguish two cases.

(1) Any two vertices in M are joined by at most one boundary. In this case the network — or graph — of the boundaries is called a *tree*. It is easy to show that there is a point P of M which belongs to exactly one boundary. Indeed, let Q be a point of M. Any sequence of boundaries starting from Q must lead to an end position P on a branch of the tree (Fig. 2.89(a)). This is so because the number of boundaries of M is finite and M has no circuits, that is closed paths of boundaries.

The point P lies on only one boundary b of M. By removing P and b from M a new map M^* is formed. M^* has $r = 1$ country, $v - 1$ vertices and $(n + 1) - 1 = n$ boundaries. By the induction hypothesis, these numbers satisfy Euler's formula:

$$(v - 1) - n + 1 = 2.$$

This equation yields

$$v - (n + 1) + 1 = 2,$$

that is, Euler's formula holds for the number of vertices, countries, and boundaries of M.

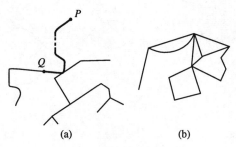

(a)   (b)

Fig. 2.89

(2) M has at least two vertices joined by at least two sequences of boundaries, s_1 and s_2 (Fig. 2.89(b)). When a boundary from one of these sequences is deleted, a new map M^* is obtained. M^* has v vertices, n boundaries, and $r - 1$ countries. According to the induction hypothesis

$$v - n + (r - 1) = 2.$$

Hence

$$v - (n + 1) + r = 2;$$

Euler's formula holds for M.

Problem 147

Let M be a map with vertices $P_1, P_2, \ldots, P_v$, boundaries $B_1, B_2, \ldots, B_e$, and countries $R_1, R_2, \ldots, R_r$, such that P_i belong to $p_i \geqslant 3$ boundaries, $i = 1, 2, \ldots, v$.

Suppose that in M the country R_j is bounded by $b_j \geqslant 6$ boundaries, $j = 1, 2, \ldots, r$. We shall prove that this assumption is false by estimating

(a) the number N_1 of pairs (P_i, B_j), where P_i is an endpoint of B_j, and

(b) the number N_2 of pairs (R_i, B_j), where B_j is a boundary of R_i.

(a) N_1 can be calculated in two ways:

$$N_1 = p_1 + p_2 + \ldots + p_v \quad \text{and} \quad N_1 = \underbrace{2 + 2 + \ldots + 2}_{e \text{ times}} = 2e.$$

Since $p_i \geqslant 3$ for $i = 1, \ldots, v$, the above equations imply that

$$2e \geqslant 3v. \tag{2.105}$$

(b) N_2 can also be calculated in two ways:

$$N_2 = b_1 + b_2 + \ldots + b_r \quad \text{and} \quad N_2 = \underbrace{2 + 2 + \ldots + 2}_{e \text{ times}} = 2e.$$

According to our assumption, $b_j \geqslant 6$. Thus

$$2e \geqslant 6r,$$

i.e.

$$e \geqslant 3r. \tag{2.106}$$

From (2.105) and (2.108) it follows that

$$2e + e \geqslant 3v + 3r,$$

that is

$$v + r \leqslant e \, .$$

This contradicts Euler's formula $v - e + r = 2$. Thus at least one country of M has fewer than six boundaries.

Problem 148

(a) Let M be a normal map, properly coloured with three colours c_1, c_2, and c_3, and let R be an arbitrary country of M with boundaries $b_1, b_2, \ldots, b_k$ listed cyclically. Denote by R_i the country separated from R by b_i, $i = 1, 2, \ldots, k$. Since M is a normal map, the countries R_i and R_{i+1} have a common border, and so do the countries R_k and R_1. Hence the colours of $R_1, R_2, \ldots, R_k$ alternate, and the colours of R_k and R_1 differ. This implies that k is even.

(b) Conversely, suppose that M is a normal map, and that every country of M has an even number of boundaries. We shall prove by induction on the number n of the countries of M that M can be properly coloured with three colours.

A normal map with three or four countries, each of which has an even number of boundaries, can be properly coloured with three colours c_1, c_2, c_3 (see Fig. 2.90).

Suppose that any map M with n or $n - 1$ countries, satisfying the conditions of our theorem, can be properly coloured with three colours.

Let M^* be a map with $n + 1$ countries, satisfying the conditions of the theorem. Since M^* is normal, it must contain a country with less than six boundaries (see Problem 147). We shall distinguish two cases.

Case 1. M^* contains a country R with two boundaries b_1 and b_2. Denote by R_1 and R_2 the countries separated from R by b_1 and b_2 respectively. R_1 and R_2 meet along the boundaries b_3 and b_4 passing through the vertices common to b_1 and b_2 (Fig. 2.91(a)).

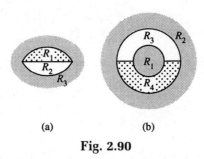

(a)   (b)

Fig. 2.90

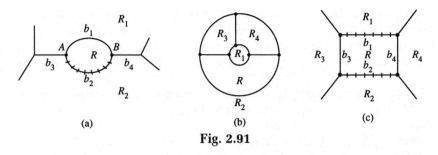

(a)   (b)   (c)

Fig. 2.91

We delete b_2 from M^*, form a new country $R_2' = R \cup R_2$, and join b_3, b_1, and b_4 into a single boundary between R_1 and R_2'. In this way a new map M is created. M consists of n countries, each of which, except R_2' and R_1, is a country of M^* with the same boundaries as in M^*. The number of boundaries of R_1 is equal to the number of its boundaries in M^* reduced by 2. An analogous statement holds for R_2'. Thus M is a normal map in which each country has an even number of boundaries. According to the induction hypothesis, we can colour M properly using three colours, say c_1, c_2, and c_3. In such a colouring let c_1 and c_2 be the colours of R_1 and R_2' respectively. We restore b_1 and R, and change the colour of R to c_3. This yields a proper colouring of M^* with three colours.

Case 2. M^* contains no country with two boundaries. Then it must contain a country R with four boundaries.

We start with the following remark. It can happen that two countries, separated from R by a pair of its opposite boundaries, meet along a common boundary (R_3 and R_4 in Fig. 2.91(b)) or coincide (R_3 and $R = R_4$ in Fig. 2.90(b)). However, under these circumstances the two remaining countries adjacent to R can have no common boundary and cannot coincide.

Denote by R_i the countries separated by boundaries b_i from R, and by k_i the number of boundaries of R_i, $i = 1, 2, 3, 4$ (Fig. 2.91(c)). We assume, without loss of generality, that R_1 and R_2 do not coincide.

Delete from M^* the boundaries b_1 and b_2, join R_1, R, and R_2 into a new country R', and delete the four vertices of R. This gives a new map M with $n - 1$ countries. The countries of M, except R', R_3, and R_4, are exactly the same, with the same boundaries, as in M^*. The countries R_3 and R_4 in M^* have $k_3 - 2$ and $k_4 - 2$ boundaries respectively. R' has $k_1 + k_2 - 4$ boundaries. Thus, M is a normal map in which each country has an even number of boundaries. According to the induction hypothesis, M can be properly coloured with three colours, say c_1, c_2, and c_3.

Suppose that in such a colouring R' has colour c_1 and R_3 has colour c_2. Then R_4 must have the same colour, c_2, as R_3. Indeed, the $k_1 - 1$ boundaries of R', which belong to R_1 in M^*, belong in the new map M to a sequence of $k_1 - 1$ pairwise adjacent countries starting with R_3 and ending with R_4 (Fig. 2.91(c)). The colours of these countries alternate between c_2 and c_3. Since $k_1 - 1$ is an odd number and the colour of the first country in the sequence is c_2, the colour of the last country in the sequence, that is of R_4, is also c_2.

Now we restore the boundaries b_1, b_2 and the regions R_1, R, and R_2. We change the colour of R to c_3, and leave the remaining colours on the map unaltered. This produces a proper colouring of M^* with three colours.

Problem 149

(a) Denote by a_i the ith term of a binary sequence of length k for $i = 1, 2, \ldots, k$. Since a_i can take any of the two values 0 and 1, the number of all binary sequences is $\underbrace{2 \cdot 2 \cdot \ldots \cdot 2}_{k \text{ times}} = 2^k$.

(b) Figure 2.92 shows a memory wheel for $k = 3$ (there are more different memory wheels for $k = 3$).

It consists of 8 digits. This is the smallest number which the wheel can contain because the number of binary sequences of length 3 is

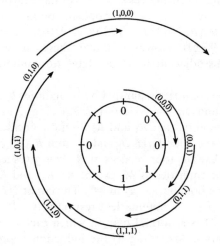

Fig. 2.92

$2^3 = 8$. All binary sequences of length 3 can be read off from the wheel starting with any sequence, say $0, 0, 0$ as follows:

$$
\begin{array}{ccc}
0 & 0 & 0 \\
0 & 0 & 1 \\
0 & 1 & 1 \\
1 & 1 & 1 \\
1 & 1 & 0 \\
1 & 0 & 1 \\
0 & 1 & 0 \\
1 & 0 & 0.
\end{array}
$$

Problem 150

Let G_k be the oriented graph, defined on p. 118.

(a) Suppose that G_k has an Eulerian circuit E. We shall construct a memory wheel W from E as follows. We travel along E, starting from an arbitrary vertex, say $s_1' = a_1 a_2 \ldots a_{k-1}$, traversing each of the oriented edges $s_1 s_2 \ldots s_{2^k}$ of G_k exactly once and returning to s_1'. The structure of G_k implies that the sequences s_i are of the form

$$
\left.
\begin{array}{rcl}
s_1 & = & a_1 a_2 \ldots a_k \\
s_2 & = & a_2 \ldots a_k a_{k+1} \\
s_3 & = & a_3 \ldots a_{k+1} a_{k+2} \\
\cdot & & \quad\ldots\ldots\ldots\ldots\ldots \\
s_{2^k-(k-1)} & = & \qquad a_{2^k-(k-1)} \ldots a_{2^k} \\
s_{2^k-(k-2)} & = & \qquad a_{2^k-(k-2)} \cdots a_1 \\
\cdot & & \qquad\ldots\ldots\ldots\ldots\ldots\ldots \\
s_{2^k} & = & \qquad\quad a_{2^k} a_1 a_2 \ldots a_{k-1}
\end{array}
\right\} \quad (2.107)
$$

Now we shall arrange the numbers $a_1, a_2, \ldots, a_{2^k}$, say clockwise, on a circle c. In this arrangement, any subset of k consecutive digits on c (listed clockwise) corresponds to a unique binary sequence of length k, and, conversely, each binary sequence of length k is represented on c by a subset of k consecutive digits. Thus the above arrangement of $a_1, \ldots, a_{2^k}$ is a memory wheel, satisfying condition (C).

(b) Conversely, let W be a memory wheel with digits $a_1, \ldots, a_{2^k}$, satisfying condition (C), and let G_k be the oriented graph, defined on p. 118. Using W, we shall construct the sequences $s_1, s_2, \ldots, s_{2^k}$, listed in (2.107). The corresponding edges $s_1, s_2, \ldots, s_{2^k}$ of G_k, taken in this order, form an Eulerian circuit of G_k (starting and ending at $s_1' = a_1 a_2 \ldots a_{k-1}$).

Problem 151

(a) For $k = 1$ the solution is obvious.

For $k = 2, 3, \ldots$ we construct the oriented graph G_k, defined on p. 118. Let $s = a_{i_1} a_{i_2} \ldots a_{i_{k-1}}$ be an arbitrary vertex of G_k. There are two edges in G_k starting at s, namely $a_{i_1} a_{i_2} \ldots a_{i_{k-1}} 0$ and $a_{i_1} a_{i_2} \ldots a_{i_{k-1}} 1$. Also there are two edges in G_k, ending at s', namely $0 a_{i_1} a_{i_2} \ldots a_{i_{k-1}}$ and $1 a_{i_1} a_{i_2} \ldots a_{i_{k-1}}$. Moreover, G_k is connected; indeed, two arbitrary, distinct vertices $a_{i_1} a_{i_2} \ldots a_{i_{k-1}}$ and $a_{j_1} a_{j_2} \ldots a_{j_{k-1}}$ are connected by the sequence of edges $a_{i_1} \ldots a_{i_{k-1}} a_{j_1}$, $a_{i_2} a_{i_3} \ldots a_{i_{k-1}} a_{j_1} a_{j_2}$, $a_{i_3} \ldots a_{i_{k-1}} a_{j_1} a_{j_2} a_{j_3} \ldots a_{i_{k-1}} a_{j_1} \ldots a_{j_{k-1}}$. Thus G_k has an Eulerian circuit.

Hence, by applying the method described in Problem 150, we can construct a memory wheel which contains each binary sequence of length k exactly once.

(b) The construction of a memory wheel for $k = 4$ is carried out in three stages.

At stage 1, the oriented graph G_4 is drawn, and an Eulerian path E of G_4 is constructed (Fig. 2.93(a)). Stage 2 consists in listing the edges

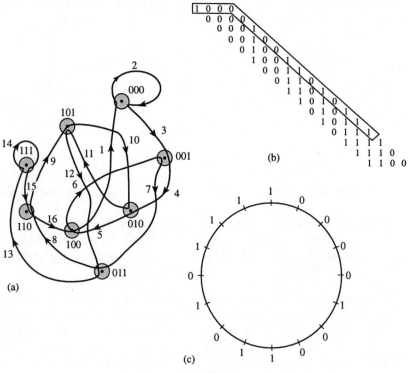

Fig. 2.93

of E consecutively (Fig. 2.93(b)). At the third stage, the digits encircled in Fig. 2.93(b) are arranged cyclically in a circle c (Fig. 2.93(c)); this arrangement of the digits is a memory wheel containing all binary sequences of length 4.

Problem 152

(a) For a given $k > 1$ we shall construct an oriented graph Γ_k such that

> the vertices of Γ_k are the sequences of length $k - 1$ with digits $0, 1, 2, \ldots, 9$,

> two vertices s_i and s_j are joined by an oriented edge leading from s_i to s_j if and only if the last $k - 1$ digits of s_i are the same as the first $k - 1$ digits of s_j.

Γ_k is a connected graph (verify). An arbitrary vertex $s_i = a_{i_1} \ldots a_{i_{k-1}}$ of Γ_k is the beginning of 10 oriented edges $a_{i_1} \ldots a_{i_{k-1}} t$ for $t = 0, 1, \ldots, 9$. Also, s_i is the endpoint of 10 oriented edges $t a_{i_1} a_{i_2} \ldots a_{i_{k-1}}$ for $t = 0, 1, \ldots, 9$. Thus Γ_k satisfies the Eulerian condition (E); we can construct an Eulerian path of Γ_k.

This implies that using the same method as in Problem 150 we can construct a memory wheel W which contains each 'denary' sequence (that is, a sequence with digits from $\{0, 1, \ldots, 9\}$ of length k exactly once.

There are 10^k denary sequences of length k. The number of digits of W is also 10^k.

(b) The study of a memory wheel of the required type is facilitated by the study of the following oriented graph D_n.

The vertices of D_n are the sequences $d_i = a_{i_1} a_{i_2}$ of distinct digits $a_{i_1} a_{i_2} \in \{1, 2, \ldots, n\}$, and two vertices $d_i = a_{i_1} a_{i_2}$ and $d_j = a_{j_1} a_{j_2}$ are joined by an oriented edge which starts at d_i if and only if $a_{i_2} = a_{j_1}$ and $a_{i_1} \neq a_{j_2}$. The edge is then denoted by $a_{i_1} a_{i_2} a_{j_2}$.

In D_n an arbitrary vertex $d_i = a_{i_1} a_{i_2}$ is the beginning of $n - 2$ edges $a_{i_1} a_{i_2} t$ and the end of $n - 2$ edges $t a_{i_1} a_{i_2}$, $t \in \{1, \ldots, n\}$ and $a_{i_1} \neq t \neq a_{i_2}$.

Thus, D_n will admit an Eulerian path if the graph is connected. This is not always the case. A counterexample, for $n = 3$, is given in Fig. 2.94.

On the other hand, for $n > 3$ any two vertices of D_n are connected. Indeed, let $s = ab$ be an arbitrary vertex of D_n, and let c, d be any two distinct digits in $\{1, 2, \ldots, n\}$, both different from a and b. The tree-diagram in Fig. 2.95 shows that the vertex ab is connected to each vertex of D_n with digits from the subset $\{a, b, c, d\}$ of $\{1, 2, \ldots, n\}$.

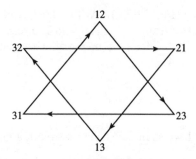

Fig. 2.94

Thus for $n > 3$ the graph D_n has an Eulerian path. Using this path we can construct a memory wheel which contains all sequences of length 3 with n digits exactly once.

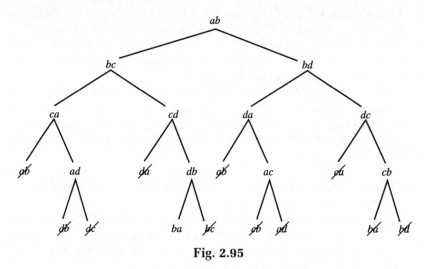

Fig. 2.95

Problem 153

(a) Figures 2.96(a)–(e) depict graphs whose vertices and edges are the vertices and edges respectively of the five Platonic solids (the regular tetrahedron, the cube, octahedron, dodecahedron, and icosahedron). These graphs admit Hamiltonian circuits; a Hamiltonian circuit for each graph is drawn in Fig. 2.96.

(b) Let D be a rhombic dodecahedron, and let G be the graph whose vertices and edges are the vertices and edges respectively of G. The number of edges issuing from different vertices of G varies. Specifically, six vertices

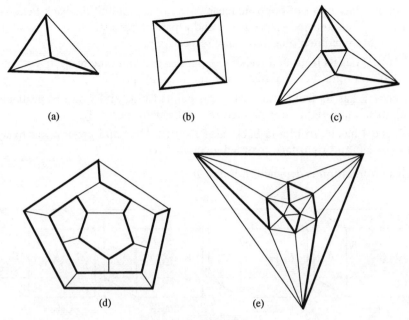

Fig. 2.96

of G are endpoints of four edges, and the remaining eight vertices are endpoints of three edges. Each edge of D has endpoints of both types.

This implies that in a Hamiltonian circuit of D the six edges of one type must alternate with the eight edges of the other type, which is impossible.

Thus D has no Hamiltonian circuit.

Problem 154

The graph in Fig. 2.97(a) contains two subgraphs with four edges labelled 1–4 (see Fig. 2.97(b) and (c)). These subgraphs provide the following information about the cubes:

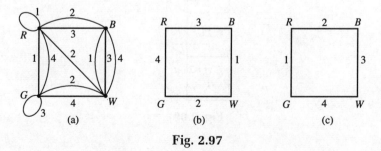

Fig. 2.97

cube 1 has a pair of opposite faces — say the front and the back faces —
painted green and red respectively, and a pair of opposite faces — say left
and right — painted blue and white respectively;

cube 2 has its front and back faces painted red and blue respectively, and
its left and right faces painted white and green respectively;

cube 3 has its front and back faces painted blue and white respectively
and its left and right faces painted red and blue respectively;

cube 4 has its front and back faces painted white and green respectively
and its left and right faces painted green and red.

This information is depicted in Fig. 2.98.

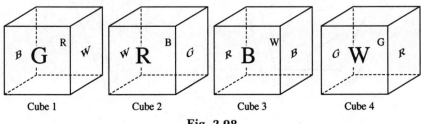

Fig. 2.98

By placing the four cubes on top of one another to form a 1 × 1 × 4 cuboid
C, so that the front, back, left, and right faces of the cubes remain parallel
to their positions in Fig. 2.98, a solution of the puzzle is obtained; Fig.
2.99 shows that each of the 4 × 1 faces of C is composed of four squares
painted in four different colours.

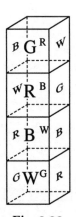

Fig. 2.99

References and suggestions for further reading

Most books in the following list are recommended for further reading. Their titles appear in *italic* type.

Chapter I

[1] M. Gardner, *More mathematical puzzles and diversions*, Penguin Books, Harmondsworth, 1976.
[2] H.O. Peitgen, H. Jürgens, and D. Saupe, *Fractals for the classroom*, parts 1 and 2, Springer, Berlin, 1992.
[3] A Rényi, Tagebuch über die Informationstheorie, Akadémiai Kiadó, Budapest, 1982.
[4] I.J. Schoenberg, Mathematical time exposures, The Mathematical Association of America, 1982.
[5] V.A. Uspensky, *Pascal's triangle and certain applications of mechanics to mathematics* (Little Mathematics Library), Mir Publications, Moscow, 1979.
[6] N. Vilenkin, *Combinatorial mathematics for recreation,* Mir Publications, Moscow, 1972.
[7] N.N. Vorobyov, *The Fibonacci numbers* (Topics in Mathematics), Heath, Boston, MA, 1966.

Chapter II

[8] W.W.R. Ball and H.S.M. Coxeter, *Mathematical recreations and essays,* (12th edn), University of Toronto Press, 1974.
[9] A.H. Beiler, *Recreations in the theory of numbers,* (2nd edn), Dover, New York, 1966.
[10] J. Cofman, *What to solve? Problems and suggestions for young mathematicians,* Oxford University Press, 1990.
See also reference 5.

Chapter III

[11] R. Honsberger, *Ingenuity in mathematics* (New Mathematical Library), The Mathematical Association of America, 1970.

[12] *Hungarian problem books II* (New Mathematical Library), The mathematical Association of America, 1963.
[13] H. Rademacher and O. Toeplitz, *The enjoyment of mathematics* (2nd edn), Princeton University Press, 1970.
[14] D.O. Shklarsky, N.N. Chentzov and I.M. Yaglom, *The USSR olympiad problem book,* Freeman, San Francisco, CA, 1962.

Chapter IV

[15] G.H. Hardy and E.M. Wright, An introduction to the theory of numbers, (5th edn), Oxford University Press, 1985.
[16] I.M. Yaglom, *Geometric transformations* (New Mathematical Library), The American Mathematical Association, 1962.
See also references 4 and 14.

Chapter V

[17] T.M. Apostol, Introduction to analytic number theory, Springer, Berlin, 1976.
[18] N.N. Vorob'ev, *Criteria for divisibility,* University of Chicago Press, 1980.

Chapter VI

[19] Dörrie, *100 great problems of elementary mathematics,* Dover, New York, 1965.
[20] N.D. Kazarinoff, *Geometric inequalities* (New Mathematical Library), The American Mathematical Association, 1961.
[21] F.G. Shleifer, On a scheme of proof of inequalities (in Russian), Matematika v shkole, Pedagogika, Moscow 1984–(6).
[22] B.L. van der Waerden, *Geometry and algebra in ancient civilizations,* Springer, Berlin, 1983.
See also references 17 and 18.

Chapter VII

[23] V.G. Boltyanski, *Hilbert's third problem,* Wiley, New York, 1978.
[24] V.G. Boltyanski, *Equivalent and equidecomposable figures,* Heath, Boston, MA, 1963.

[25] H.S.M. Coxeter, Twelve geometric essays, Southern Illinois University Press, 1968.
[26] H.S.M. Coxeter and S.L. Greitzer, *Geometry revisited* (New mathematical Library, The American Mathematical Association, 1965.
[27] H. Eves, *An introduction to the history of mathematics*, Holt, Rinehart and Winston, New York, 1964.
[28] H. Meschkowski, *Unsolved and unsolvable problems in geometry*, Miver & Boyd, Edinburgh, 1966.
[29] I.F. Sharygin, *Problems on Geometry: stereometry* (in Russian), Bibliotečka Kvant, Nauka, Moscow, 1984.
[30] A. Wassermann, *The 13 spheres problem*, Eureka, Cambridge, 1978.
See also reference 19.

Chapter VIII

[31] H.S.M. Coxeter, Regular polytopes, (3rd edn), Dover, New York, 1973.
[32] J. Dieudonné, Linear algebra and geometry, Kershaw, London, 1969.

Chapter IX

[33] W.W.R. Ball and H.S.M. Coxeter, *Mathematical recreations and essays* (12th edn), University of Toronto Press, 1974.
[34] P. Bossert, *You can do the cube,* Puffin Books, Harmondsworth, 1981.
[35] V. Dubrovski, Mathematics of the magic cube (in Russian), Kvant, Nauchno-populyarnij fiziko-matematicheskij zhurnal, Nauka, Moscow, No. 8, 1982.
[36] O. Ore, *Graphs and their uses* (New Mathematical Library), The Mathematical Association of America, 1963.
See also reference 8.

Chapter X

[37] E.B. Dynkin and V.A. Uspenskii, *Multicolor problems,* Heath, Boston, MA, 1963.
[38] L.I. Golovina and I.M. Yaglom, *Induction in geometry* (Little Mathematics Library), Mir Publications, Moscow, 1979.
[39] G. Pólya, R.E. Tarjan and D.R. Woods, Notes on introductory combinatorics, Birkhäuser, Boston, MA, 1983.
[40] H.J. Ryser, Combinatorial mathematics (The Carus Mathematical Monographs), No. 14, The Mathematical Association of America, 1965.
See also reference 35.

Index